ADVENTURES
IN
MEMORY

FOREWORD BY **SAM KEAN**

HILDE ØSTBY & **YLVA ØSTBY**

TRANSLATION BY **MARIANNE LINDVALL**

ADVENTURES

IN MEMORY

The Science and Secrets of Remembering and Forgetting

GREYSTONE BOOKS

Vancouver/Berkeley

18 19 20 21 22   5 4 3 2 1

Greystone Books Ltd.
greystonebooks.com

Cataloguing data available from Library and Archives Canada
ISBN 978-1-77164-347-4 (cloth)
ISBN 978-1-77164-345-0 (epub)

Editing by Jennifer Croll
Copy editing by Dawn Loewen
Jacket and text design by Nayeli Jimenez
Jacket illustration by H. L. Todd, U.S. National Museum
Printed and bound in Canada on ancient-forest-friendly paper by Friesens

Greystone Books gratefully acknowledges the Musqueam, Squamish, and Tsleil-Waututh peoples on whose land our office is located.

Greystone Books thanks the Canada Council for the Arts, the British Columbia Arts Council, the Province of British Columbia through the Book Publishing Tax Credit, and the Government of Canada for supporting our publishing activities.

Canadä

CONTENTS

FOREWORD

Sam Kean

I N ONE OF Plutarch's *Lives*, he mentions an old philosophical conundrum now called the Ship of Theseus Paradox. After slaying the minotaur in the labyrinth, the Greek hero Theseus sails home, and the people of Athens are so overjoyed that they preserve his ship as a memorial. Being a material object, however, the ship shows some wear and tear as the years pass, and every so often the caretakers have to replace a piece of it—a board here, a plank there. Eventually, after several centuries, they've replaced every last original piece of wood. So is it still the same ship now that it was at the beginning?

You can ask a similar question about our own bodies. We're three-fourths water, molecules of which cycle into and out of us in a constant flow, never sticking around long. And because our DNA and other biomolecules break down, we swap in new parts all the time to repair them. Skin cells get recycled every few weeks, blood cells every few months, liver cells every two years. Even bone experiences constant turnover. Over the course of a decade or so, every last atom in your body gets replaced. So are you still *you* at the end?

Most of us feel, intuitively, that we do remain the same person from decade to decade. But how can that be if our bodies aren't the same? Why do we feel this continuity so strongly? One big reason is because of our memories. The bio-bits come and go, but the pattern of information coursing through our brains remains largely consistent. In a fundamental way, then, we *are* our memories.

Yet, few of us really understand how memory works. We rely on misleading analogies or folk theories, or we simply remain oblivious and take it all for granted: memories bubble up so easily inside us, so effortlessly, that we rarely pause to consider just what a miracle they are.

Well, no more. After reading *Adventures in Memory*, you'll fully appreciate what a complex, beautiful, and intricate thing memory is. Indeed, in the hands of Hilde and Ylva Østby, memory is more than a simple repository of our selves. It's a creative force—something dynamic that actively shapes our thinking. It has all the rich, layered complexity of human beings themselves.

What makes the book unique is the double-barreled perspective the Østby sisters bring. Unlike, say, electrons or black holes, memory is both an objective and a subjective phenomenon. We need to understand the neurotransmitters and electrical firing patterns involved, but memory is a *lived* thing too. We need both sides—in other words, the literary and the scientific—to make sense of memory. It's therefore reassuring to know that Hilde Østby is a novelist and her sister Ylva a neuroscientist. Imagine if another sibling pair—William James the famous psychologist and Henry James the brilliant novelist—had combined forces, and you can see what a valuable perspective this combination provides.

We usually think about science as ultra-rational, but it's an intensely human activity as well. And it's especially important to see that human side when studying memory, since it does influence our sense of self so profoundly. In exploring how memory works, then, *Adventures in Memory* is really a dive into your innermost self—the most intimate aspects of your being. And when you surface from this dive, you'll never see the world, or your own self, in quite the same way again.

SAM KEAN, author of *Caesar's Last Breath* and *The Tale of the Dueling Neurosurgeons*

THE SEA MONSTER

Or: The discovery of the hippocampus

Your memory is a monster; *you* forget—*it* doesn't. It simply
files things away. It keeps things for you, or hides things
from you—and summons them to your recall with a will of its
own. You think you have a memory; but it has you!

JOHN IRVING, *A Prayer for Owen Meany*

A T THE BOTTOM of the ocean, tail curled around seagrass, the male seahorse sways back and forth in the
current. He may be tiny and mysterious, but no ocean
creature compares to him. The only male in the animal
kingdom to become pregnant, he stands on guard, carrying his eggs in his pouch until they hatch and the fry swim
away into the open sea.

But let's back up: this isn't a book about seahorses. To
find our real subject, we must rise out of the depths and
journey back 450 years.

The year is 1564. We're in Bologna, Italy, a city full of elegant brick buildings and shady, vine-covered walkways. Here, at the world's first proper university, Dr. Julius Caesar Arantius bends over a beautiful object. Well, beautiful might be an exaggeration, if you're not already deeply, passionately involved in its study. It's a human brain. Rather gray and unassuming, and on loan from a nearby mortuary. Students surround the doctor, clustered on benches throughout the theater, following his work intently, as though he and the organ in front of him are the two leads in a drama. Arantius leans over the brain and slices through its outer layers, studying each fraction of an inch with extreme interest, hoping to understand what it does. His disregard for religious authority is clear in the gusto with which he approaches his dissection because, until shortly before that time, the scientific study of human corpses had been strictly forbidden.

The doctor cuts further into the object, examining what's inside. And then, deep within the brain, buried in the temporal lobe, he finds something very interesting. Something small, curled up into itself. It looks, he thinks, a bit like a silkworm. The upper classes of the Italian Renaissance loved silk, a luxurious and exotic fabric that arrived in Venice via the Silk Road from China; by extension, they loved silkworms too. Intrigued, Arantius looks closer, making some careful cuts, and pries the little worm loose, liberating it from the rest of the brain.

This is the moment at which modern memory research was born, the precise moment that memory, as a concept, moved from the mythological world into the physical one. However, back then, on that particular day in sixteenth-century Bologna, life goes on in the markets as usual; people carry wine and truffles and pasta below the city's famous pergolas and ancient red brick towers, oblivious to the hugely important discovery in their midst.

Arantius turns over what he has dug out of the brain and places it on the table before him, considering what it might be. That's it! Rather than a silkworm, perhaps it is a tiny seahorse? Yes, indeed. With its head nodding forward and its tail curling up, it does look like a seahorse, the tiny distinctive fish living in shallow ocean waters between the tropics and England. And so he names it: *hippocampus*, meaning "horse sea monster" in Latin. It also shares its name with a mythological creature—half horse, half fish—said to wreak havoc in the waters around ancient Greece.

By the light of a tallow candle perched on an autopsy table, Julius Caesar Arantius couldn't tell what this little part of the brain actually did. All he could do was give it a name. Hundreds of years passed before we fully understood the significance of what this Italian doctor held in his hands, and you might guess that it has something to do with memory. After all, memory is the subject of this book.

The world beneath the sea and the one in our brain are profoundly different, of course, but there are many similarities between the seahorse and the hippocampus. Just as the male seahorse carries his eggs in his pouch until it's safe for the fry to be on their own, the seahorse of the *brain* also carries something: our memories. It watches over them and nurtures them until they are strong enough to make it on their own. The hippocampus is the womb that carries our memories.

No one knew how crucial the hippocampus was to memory until 1953, but there was endless speculation about where memories were stored in the brain. One popular early belief was that our thoughts flowed through the liquid inside our skulls, but that theory was long gone by 1953. By then, the prevailing thought was that memories were created and stored throughout the brain. But then something happened to sink this theory once and for all, an incident that was tragic for one man, fortunate for the rest of us. An unsuccessful experimental surgery was the key to understanding Julius Caesar Arantius's earlier discovery.

IF HENRY MOLAISON had lived in modern times, his treatment would have been very different. Henry suffered from severe epilepsy, and several times a day—or sometimes several times an hour—he had small absence seizures (also known as petit mal seizures), in which he would black out for a few seconds at a time. At least once a week he'd suffer a major convulsive seizure (or a grand mal seizure), in which he'd completely lose consciousness and his body would shake violently for several minutes. The medicines he was prescribed only made things worse and resulted in more seizures. In 1953, at age twenty-seven, he sought treatment from a surgeon, William Beecher Scoville, who proposed an operation unthinkable by today's standards.

Dr. Scoville didn't have the benefit of hindsight. He was inspired by reports of a Canadian surgeon who'd removed the hippocampus on one side in several patients in order to cure their epilepsy, and he believed that if he removed the hippocampi from both sides of Henry's brain, then the treatment would be twice as effective. Henry listened to his doctor. After a lifetime of crippling epilepsy, he was desperate, and he agreed to the operation. Unwittingly, Henry Molaison had signed up to become the most important subject in the history of memory research.

When Henry woke up after the surgery, doctors found that he had no memory of the last two or three years; in fact, he couldn't retain anything beyond what was present within his short-term memory. The nurses had to show him the way to the bathroom every time he needed to go. He had to be constantly reminded of where he was, because he forgot as soon as he thought of something else. He had lost his ability to form new memories.

For the remaining fifty-five years of his life, Henry lived literally in the moment. He couldn't remember what he'd done half an hour ago, or the joke he'd told a minute earlier. He couldn't remember what he'd eaten for lunch and had no idea how old he really was, until he looked in the mirror and saw gray hairs. He had to guess what season it was when looking out the window. Since he couldn't remember new information, he couldn't manage his money, diet, or household chores, so he lived at home with his parents. In spite of this, he was usually calm and content. But sometimes, things would upset him—like the death of his father.

Each morning, Henry woke up with no recollection of his grief, but every time he rose, he made an alarming discovery: the valuable weapons collection normally hanging on the wall was missing. He understood there was something wrong—the weapons were gone—and concluded that the house had been burgled and the weapons stolen. The truth was that his uncle had inherited the collection. But there was no use explaining that this was due to his father's death, as the next morning he would conclude, all over again, that he'd been burgled. Finally, his uncle had to return the weapons collection. Eventually, Henry appeared to get used to the fact that his father didn't come home anymore and, to some degree, understood that he was gone.

Scoville's surgery on Henry was an experiment, but no one at the time could have anticipated the consequences. In fact, Scoville had already performed the operation on dozens of other patients. None had shown any obvious signs of memory loss. But there was a catch: every patient in this group had been acutely schizophrenic, paranoid, or psychotic. Their behavior was already abnormal, so any memory problems were blamed on their psychosis. Incidentally, they were no less schizophrenic after the surgery. But this was the era when lobotomies were in fashion, and Scoville believed he could improve this procedure by removing the hippocampi rather than following the classic approach of removing parts of the brain's frontal lobe. Exactly why he believed this is another story. *Our* story centers on the consequences of his famous surgery on Henry Molaison. But there were also consequences for Scoville, who was deeply concerned about the results of the surgery. In a scientific paper he cowrote with Canadian neuropsychologist Brenda Milner in 1957, he confessed to his mistake. In the years after the article was published, Milner endeavored to find out more about how Henry's memory was damaged. She believed that together, she and Henry would be able to explain to the world how human memory functions.

What could be learned about memory by studying Henry Molaison? Simply talking to him revealed some basics about the structure of memory; he was quite capable of sticking to the topic of conversation, as long as his thoughts didn't wander and he didn't become distracted by something around him. This meant that he had normal *short-term* memory. Short-term memory is what we remember *in the moment*. Before our experiences become *permanent* memories, they spend time in short-term memory. When we look up a phone number, we remember the number for a short while before we dial it. The same happens when we learn a new word or somebody's name. These things remain in our memories for no more than a few seconds, or as long as we keep thinking about them. Sometimes, items that pass through short-term memory are picked up for long-term storing. In this case, all that remained was Henry's short-term memory, but he learned how to use it in ingenious ways. During one study, examining his ability to perceive time, a researcher told him that she would leave the room, and that when she returned, she would ask him how long he thought she'd been gone. He suspected that he wouldn't be able to do this, so he did something clever; he looked at the wall clock (something the researcher hadn't noticed) and memorized the time by silently repeating it to himself, over and over again, until she returned. When she came back, he looked at the clock again to calculate how long she'd been gone. Since he'd focused on this one task during the test, he was able to keep the information in his short-term memory. Henry knew that he was participating in an experiment. But he couldn't remember the researcher or her name.

Luckily, Henry enjoyed mental challenges; he always had a crossword with him and happily solved puzzles, which made him a willing participant in Brenda Milner's experiments. In one example, she gave him a small maze, in which he had to negotiate steps through a grid to find an exit. After 226 attempts, he still couldn't do it. But since he had no recollection of all his previous attempts, he happily carried on trying.

Another time, Milner asked him to draw a star while looking only at a reflection of his hand in a mirror. For anyone, this is a difficult task, because when you're watching a mirror image, you tend to move the pencil in the wrong direction when you get to a corner of the star. But with practice you can improve. It's a way of learning, of remembering, that helps you perform a task better the next time. But unlike remembering *events* you have experienced, or figuring out a maze, this kind of memory doesn't involve conscious thought. It's like riding a bicycle; you never remember that you have to move your feet a certain way, or tilt your body to keep your balance; it becomes second nature. Each time Henry repeated the mirror drawing task, he too improved. As with people with intact hippo-campi, he eventually mastered the mirror star test. His final, almost perfectly drawn star surprised him, because he had no knowledge of the previous attempts that had helped him to gradually improve.

"That's strange. I thought it would be difficult, but it looks as though I've done it rather well," he said, astounded.

Brenda Milner was also astounded. She'd made a crucial discovery about long-term memory: that it consists of different, separate storage areas. In learning tasks that are not based on conscious memories, but rather on procedural memories (the body's memory of how to do something), the hippocampus is not involved. If it were, Henry wouldn't have done so well.

IN LATER YEARS, Suzanne Corkin, a student of Brenda Milner's who became a neuroscientist, took over the work of researching Henry Molaison's memory. Corkin and Henry's partnership lasted for more than forty years, and in a way continued past his death. But though the two saw each other frequently, and to her he was like an old friend, *she* was new to him every time they met. When she asked him if he knew who she was, he replied that there was something familiar about her. He would guess that she was perhaps an old schoolmate. He may have wanted to be polite, but perhaps there was a remnant of something like a memory in his brain, which gave him a feeling of recognition, without knowing where it came from.

Henry reaffirmed that we possess a short-term memory, something he still had, and a long-term memory, which he had only half of—the part involving unconscious learning, otherwise known as *procedural memory*. What he was missing was the ability to store memories that can be consciously recalled: facts about the world and himself, called *semantic memories*, and all the experiences that would normally become part of his personal memory album, called *episodic memories*.

The modern theory of memory—based in part on Henry Molaison—suggests that previously stored memories are separate from new memories waiting to be let in. Henry did have memories from before his surgery. He remembered who he was and where he came from. He remembered events from his childhood and youth. But the three years leading up to the surgery were completely gone. This meant that memories could not be stored in the hippocampus, or at least not *only* there. Anyway, it is unlikely that there could be room for all of life's experiences in such a tiny, fragile structure deep inside our brain. The role of the hippocampus must be to hold on to memories while they are maturing, before they are properly stored elsewhere in the brain—in the cerebral cortex, the outer layer enveloping the brain. It's logical to think this process may take about three years, since Henry couldn't remember the three years prior to his unfortunate surgery.

As Henry was living life from one minute to the next, in the safety of his mother's house, the continual experiments turned him into something of a memory celebrity. Fortunately, researchers kept his identity hidden until after his death. If they hadn't, he would have been vulnerable to overenthusiastic researchers and journalists. He was known only by his initials, and to this day, memory researchers around the world refer to him only as H.M.

Henry contributed his life to research—or at least the memories of his life. He took part in one experiment after the other, so researchers could document how memory works. Although he remembered very little after the surgery, he had memories of conversations with his doctor from several years prior to the surgery. This meant he understood that something had gone wrong—maybe with the surgery. This is why he repeatedly told the researchers that he wanted to help prevent the same thing from happening to others. "It's a funny thing—you just live and learn. I'm living, and you're learning," he said.

Another important consequence of the research on Henry was that no one *was* ever operated on in the same way again. Scoville quit removing both hippocampi from his patients, whether they suffered from epilepsy or schizophrenia. Surgery to cure epilepsy did continue, however, and is still carried out today. If a patient has a certain type of epilepsy, originating in the area of the hippocampus, it can sometimes be remedied by removing one of the seahorses. The other one is left intact, so new memories still have at least one entrance into long-term memory.

For those of us whose brains are pretty much intact, it's easy to take memory for granted. It's easy to think, "I'm sure I'll remember this, I won't need to write it down." All the special moments in our lives will remain with us as memories, won't they? We like to imagine memory as a hard drive filled with film clips from our lives that we can watch whenever we want to. That's not how it works, though. When we're driving to the store or sitting around the dinner table with good friends and family, how can we be sure those precise moments will be remembered? Will those memories be useful or important in the future? Our memories do take good care of certain moments, of course: birthdays, weddings, a first kiss, the first time we score in soccer. But all the other moments—what happens to them? We spring-clean our brains now and then, throw out the clutter, and keep some things for safekeeping. This is a good thing, because if we had to remember every single moment of our lives, we wouldn't be able to do much more than reminisce. When would we have time to live?

Some of us, however, store more than others: meet Solomon, the man who was unable to forget anything at all!

Solomon Shereshevsky worked as a Russian newspaper journalist in the 1920s. There, he annoyed his editor by never taking notes when he was given an assignment. The editor distributed the stories for the day, and while the other reporters eagerly wrote down what they needed to know to get to work, Solomon just sat there, as if he couldn't care less.

"Haven't you got anything of what I said?" the chief editor would ask.

But Solomon had gotten all of it: every address mentioned, every name, what the issue was. He could repeat it all back to his editor, every last detail. "Isn't this the way it is for everyone?" he thought. He found it odd that others had to take notes. To him, it was natural that everything he heard, he remembered. Solomon's editor sent him to see an expert. In the office of neuropsychologist Alexander Luria, Solomon was—like Henry Molaison—exposed to a battery of tests. How much was it possible for a human being to remember?

An almost limitless amount, as it turned out. At least it was difficult to find limits for Solomon's memory. The psychologist showed him long lists of nonsense words and he could regurgitate them in perfect order, even backward and diagonally. He memorized poetry in other languages, tables of numbers, and advanced math in the blink of an eye. When Solomon met Luria again, seventeen years later, he could still repeat the same lists he had seen that time many years ago.

Solomon eventually quit his job at the newspaper and launched a new career as a mnemonist, a memory artist. He appeared onstage and memorized endless lists of numbers or words provided by the audience. Then he repeated them perfectly, to everyone's amazement. But contrary to what you may think, an amazing memory—the kind so good we dream of having it ourselves—didn't make Solomon rich, nor did it make him powerful or particularly happy. He jumped from job to job and finally died alone in 1958, without friends or family by his side.

Solomon Shereshevsky's astonishing memory was partly due to something called synesthesia. This is a condition in which all sensations are accompanied by another sensation, such as sight, sound, smell, or taste. Solomon suffered from an extreme form of synesthesia. Everything he experienced was accompanied by impressions of bright colors, strong tastes, or special images. Hearing certain words would conjure distinct pictures, even tastes and smells. Certain voices evoked strong visual impressions. Once, when he was buying an ice cream at a kiosk, he recoiled in disgust because the seller's voice made him see a billowing storm of black coal and ashes. These profound sensations made his memories latch on far stronger than a regular person's. It was said that he couldn't get rid of a memory—not even a meaningless list of numbers—unless he made a conscious effort to remove it.

Solomon was special. Almost no one remembers things as well as he did. Compared with his, the memory of an average person is a mere joke. But would you really want to be able to remember not only your parents' phone number and the bus schedule from elementary school, but all phone numbers and bus schedules you have ever encountered?

Exactly fifty years after Solomon passed away, eighty-two-year-old Henry Molaison also died. The difference between these two exceptional men is not only that one of them had a vast trove of memories while the other couldn't remember a thing. The fifty years between them also made a difference in terms of how their memories were researched. While we know a lot about Henry's brain, we know nothing about Solomon's. We don't know if he had an extra large—or in any other way different—hippocampus. Meanwhile, Henry Molaison is still, even after his death, contributing to science. In his will, he bequeathed his brain to research, and the researcher who worked most closely with him for the last forty years of his life, neuroscientist Suzanne Corkin, planned to give her subject an afterlife in her field. After Henry's death on December 2, 2008, Corkin worked together with a large team of physicians and researchers to make his brain work for posterity. First, researchers at Harvard scanned the brain with a magnetic resonance imaging (MRI) machine in Boston. Then, Corkin placed Henry's brain in a cooler and handed it over to brain researcher Jacopo Annese, who took it on a plane bound for San Diego. Annese's Brain Observatory stores the donated brains of deceased people so they can be used in various avenues of research, including on Alzheimer's and normal aging. There, Annese's team was ready and waiting to cut Henry's brain into slices, thin as strands of hair. "We believe that the enormous attention that was devoted to patient H.M. when he was living and generously served as a keen research subject ought to be matched by a similarly involved study of his brain," Jacopo Annese said.

Henry's brain needed special attention. No other brain at the Brain Observatory had received as much scientific attention as his. The team photographed every single one of the 2,401 slices of Henry's brain and stored them both in formaldehyde and as digital files. They spent fifty-three hours doing so, and Annese didn't sleep until he was sure that all the pieces of this exceptional brain were securely preserved. Thanks to his work, researchers can now study the exact location where Scoville made his mistake and speculate about which of the remaining areas, near the hippocampus, helped Henry remember the few things which he occasionally and surprisingly did remember. In May 2016, Corkin passed away at age seventy-nine, and her brain is now in the safekeeping of other brain researchers. It contains no unusual surgical scars but houses decades of memories of her special contributions to research.

Henry Molaison's legacy was an entirely new field of research. Now the hippocampus has a definite place in our memory. And during the past fifty years, memory research has become more and more focused on mapping memories all the way down to the cellular level.

"I believe that we will achieve the goal of explaining memory in the brain within my lifetime," says one of the leading memory researchers in the world, Eleanor Maguire, professor at University College London and Wellcome Trust Centre for Neuroimaging. Her research, which focuses on the hippocampus, has allowed her to "see" memories. In one experiment, she told test subjects to think of a certain memory while she watched, through an MRI machine, the patterns that lit up their hippocampus. When they thought of other specific memories, different patterns became visible.

"Your experiences are taken into the brain. After that, the experience is taken apart and stored away in little pieces in the brain's neocortex. Every time you recollect it, it is brought back to life. The hippocampus is critical to reconstructing the memory in your mind's eye, enabling you to relive it once more," says Maguire.

Memory research is also, in a sense, a process whereby small pieces are assembled into a larger puzzle. Memories cannot be seen, per se. No one can retrieve a memory and put it under the microscope. That's also why it took so long for memory to move from being strictly a philosophical and literary topic to being the object of scientific examination. Psychology is a relatively new academic discipline, and so the scientific study of memory has a shorter history than that of many other subjects. But when memory researchers started piecing together human memory, they gave us a picture of an amazing inner world. They worked tirelessly with lists of words, meaningless shapes, staged bank robberies, life stories, puppet shows, and strings of numbers, all to reveal the truth about memory by using the brains of the people who volunteered to be guinea pigs.

Some of you will probably argue that it's meaningless to measure something so abstract, a thing that exists only for the individual who owns the memory. How will we be able to reduce the evocative descriptions of memories in Marcel Proust's seven volumes of *In Search of Lost Time* into figures and scientific graphs?

To capture unique human experiences and turn them into science, isn't that a paradox? Like putting a seahorse in a glass of formaldehyde hoping to preserve its beauty and essence forever?

There are, however, many good arguments for why memory research is necessary. Turning memory into something concrete and measurable helps us compare memory in the healthy and in those with diseases, and it can help people with memory problems. It contributes to our understanding of how the brain works, which in turn may help find the solution to major medical issues of our time, such as Alzheimer's, epilepsy, and depression.

Some 140 years of measuring memory has not solved all its riddles—far from it. Disagreements come and go on the memory battleground. One longstanding dispute is referred to as the "memory wars." One side maintains that, in extreme situations, memory will behave differently, producing things like repression and dissociation. The other side maintains that memory always behaves the same way, only much more strongly in extreme situations. Another hot issue is the possibility of memory training: Is it like strengthening a muscle—that is, it gets better with repetition—or can you use strategies and techniques to improve existing ability? And what exactly *is* memory? Even this is being debated in minute, technical detail in scientific papers—debates punctuated by indignant letters to the editors of scientific journals—while researchers try to gain ground in the scientific community. It's almost like an election campaign in slow motion, or a TV debate spread out over fifty or a hundred years.

There's even discord over the hippocampus. Two camps face each other. One rigidly believes that the role of the hippocampus is simply to consolidate memories into the rest of the brain. As time passes—sometimes with the help of a good night's sleep—memories attach themselves to more robust cortical networks, while the seahorse slowly and carefully lets go of the memories it has been tending. The other camp argues that this is too simple. This group adamantly insists that the seahorse holds on to memories, especially the personal, vivid sort that we remember in something resembling a personal memory theater, at the same time as they are also stored deeper in the cortex. Every time we recall a memory, they say, the hippocampus is involved and "overwrites" the original memory, each time with a slightly new interpretation or reconstruction.

In the same way as the seahorse's ocean ecosystem is important to understanding its existence, the hippocampus's brain ecosystem is important to understanding how memory is kept and recalled. In the past few years, people have been paying more attention to how the hippocampus interacts with the rest of the brain. Memories play out in physical networks, where different parts of the brain move as in a synchronized dance. It's visible with modern MRI methods. William James, one of the fathers of psychology, understood it already in 1890:

"What memory goes with is, on the contrary, a very complex representation, that of the fact to be recalled plus its associates, the whole forming one object,... known in one integral pulse of consciousness... and demanding probably a vastly more intricate brain-process than that on which any simple sensorial image depends."

In other words, each memory consists of different bits and pieces brought together in one unified wave of consciousness. Each part of the memory originates in a different part of the brain, where it first made a sensory impact. To make the whole thing feel like one experience, one unique memory, requires intricate brain interaction. William James didn't know exactly how this worked, but thinking of memory and mind the way he did in the 1890s was remarkable. When James was alive, people thought of each memory as a unit, a copy of reality, like something that could be pulled out of a folder in a filing cabinet. That the key to understanding memory was the seahorse—slowly swaying in rhythm with the sensory areas and the emotion and awareness centers of the brain—wouldn't be discovered for another hundred years. Just a few years before James's armchair observations, researchers had discovered how neurons are connected to each other with a slight gap in between them called a synapse: the so-called *neuron doctrine*. From that discovery to today's brain research, where we can virtually watch memories come to life in the brain, has been a long journey.

We can all benefit from making that journey and learning more about our memories. A small seahorse turned out to be the key to many of the brain's mysteries. When Julius Caesar Arantius named it the hippocampus, it probably was not solely due to its appearance. Seahorses, like silkworms, were special and somewhat mysterious during the Italian Renaissance. When an event is special and unique, it helps the hippocampus hold on to it as a memory. We know that now, but Arantius could not have known that about the tiny part of the brain he had discovered. He just wanted his discovery to be noticed—and remembered.

DIVING FOR SEAHORSES
IN FEBRUARY

Or: Where do the memories go?

Memories have huge staying power, but like dreams,
they thrive in the dark, surviving for decades in
the deep waters of our minds like shipwrecks on the sea
bed. Hauling them into the daylight can be risky.
J.G. BALLARD, *"Look Back at Empire,"*
The Guardian, *March 4, 2006*

BEYOND THE PIER at Gylte Diving Center, an hour's
drive out of Oslo, Norway, there are more than forty
different types of marine slugs (nudibranchs). They
come in all colors, from dark purple to transparent white.
Their bodies are covered with tentacles with small stars
at the tip, or are decorated with pink fringes like a Disney
character from the 1950s. They stretch orange fingertips

toward the shiny ceiling of the water's surface or defiantly pull their luminescent, light-green feelers into their bodies. They slither around in clouds of glittery particles that swirl around the water here by the pier.

The water is only forty degrees Fahrenheit. Farther into the fjord, we've seen ice floes bobbing up and down at the water's edge. Soon, the slugs in the water will be joined by ten black-clad men chasing the seahorse's secret. The divers' flippers thump against the pier as they hobble like penguins toward the sea, then swirl up clouds of particles as the divers slowly sink to a depth of fifty feet. From our vantage point on the pier, we can see bubbles on the dark surface of the water, revealing where the divers are. The seahorses they are looking for are not in the water—we are, after all, in the Oslo Fjord. No, they are hidden beneath their tight diving hoods. The divers have plunged into the ice-cold February water to find out what goes on in the hippocampus. They're hunting for memory.

Together we are going to find out how memories behave when they enter our minds. Researching memory is, in a way, quite similar to diving. Our divers are about to break the surface and descend into the depths of memory itself. The only sign that there are memories below the surface is what rises and bursts, like the divers' bubbles breaking the surface of the water.

The experiment we are re-creating, famous in memory research, was first conducted in 1975 off the coast of Scotland. Memory researchers Duncan Godden and Alan Baddeley decided to test a popular idea, that you can remember something better when you return to the place where it happened. You know, like in crime novels, where the detective remembers an important detail when he returns to the scene of the crime. It's a simple theory: when we are in the same environment as we were when an event took place, the memory of it will come streaming back, whether we want it to or not.

Are memories easier to recall at the location where we first encountered them? How and where do they find a permanent place in our brain? To properly test this, Godden and Baddeley constructed an experiment in which people had to perform a task in two different environments, on land and underwater. Their assignment was to memorize lists of words either on the pier or twenty feet deep in the water, and later recall the lists either on the pier or in the water. One list was to be learned on the pier and recalled on the pier. After some time, a second list was to be learned underwater and recalled on the pier; a third list was learned underwater and recalled underwater; and a fourth list of words was learned on the pier and recalled underwater. The researchers anticipated that everything going on in the water—the cold and wet environment, breathing through masks, and so on—would make the divers remember less than they would on the pier. Theoretically, it should also be harder to learn something underwater as opposed to on land, given that the pressure and the mixture of gases the divers breathe would make it more difficult to focus.

On this cold February morning in 2016, when we send our divers into the Oslo Fjord, it's the first time anyone has repeated Baddeley and Godden's experiment in seawater—some have re-created it in a swimming pool, but we all know that's not the same. Will these ten men—thirty to fifty-one years of age—show the same results as in the legendary British experiment?

"Now I can tell you exactly where I have been underwater, after many thousands of dives. I could not do that before," says hobby diver Tine Kinn Kvamme, the experiment photographer. The lack of oxygen underwater, along with the stressful experience, means that people's brains function differently from how they usually do.

"When people first start to dive, few remember anything at all, nor can they report what happened underwater. First-time divers are asked to write their names backward underwater. Often, they will write things like 'backward,' or they will turn around only one letter in their name. If you ask them how many wheels a cow has, they'll answer four," she says.

Ordinarily, memories reside within a large brain network. When memories enter our brain, they attach themselves to similar memories: ones from the same environment, or that involve the same feeling, the same music, or the same significant moment in history. Memories seldom swim around without connections, like a lonesome fish. Instead, they are caught in a fishing net full of other memories. When you want to recall a memory, you have a greater chance at catching it if you scoop up the other memories around it. When you pull in the net, it's full of memories, and you can keep hauling it in until you find the memory you're after.

Would memory still work this way in a stressful situation, with subjects who had to deal with diving equipment and other distractions? Would context help the divers remember what they learned underwater when they're also asked to remember it underwater?

The experiment in 1975 showed the expected results: the word lists memorized underwater were also better remembered underwater, and the lists memorized on land were better remembered in a dry environment. We anticipate the same from our divers, but we don't want their expectations to influence the results, so we haven't told them what happened in the original experiment.

The atmosphere is tense at Gylte Diving Center. We are not re-creating this classic psychological experiment just for fun: results from psychological experiments are not always reliable. A great deal can happen by coincidence, and it's often only the results that confirm the hypothesis that are reported, while those researchers who find opposite results tuck them away in a drawer, ashamed and disappointed. When a team of researchers took on the task of re-creating one hundred experiments from different areas of psychology, only thirty-six were successful. The diving experiment was not among the hundred re-creations—but it's having its time today, an ice-cold and rainy February day in Drøbak.

Throughout history, philosophers and authors have asked themselves what memory is, how we learn and remember things, and what makes a memory reappear. At risk of offending an entire professional group: in many ways, we can call the philosophers of ancient times neuropsychologists, because they observed and tried to understand how the brain works without having access to today's research methods. The million-dollar question that everyone is trying to answer is where in our brains our memories actually end up, and how it is possible for all our experiences to consolidate into a pink mass of brain cells and blood vessels. In 350 BCE, in *De Memoria et Reminiscentia* (*On Memory and Reminiscence*), Aristotle compared the memory process to making an impression in a wax seal. But exactly how the experiences turned into memories, he couldn't say.

By studying the divers at Gylte, we may not be able to see their brains etching words into wax seals, but we can observe how memories connect and become dependent on each other. Context-dependent memory tells us something very basic about how memories are stored. How much you know in a broad sense determines what you understand of the new things you learn. Your understanding of your new experiences depends on your prior experiences. This network of knowledge creates context for the new learnings—they get caught in the fishing net, if you will. When you know what the French Revolution was all about, it's easier to understand the Russian Revolution, and when you have gained insight into Russian Communism, it shines a new light on the French republics, and so on. When our divers eventually resurface—ice-cold faces and eager eyes— and hand us their notebooks, filled with all they remember from a list of twenty-five short nonsense words, we will see with our own eyes how their brains have worked, linking words and seaweed and cold water together into the same network. But we're still standing on the pier, while the February cold eats its way into our woolen underwear. It's anything but magical.

By contrast, during the Renaissance, in the 1500s and 1600s, many viewed memory as something magical. At the time, magicians and alchemists not only tried to make gold, but first and foremost used rituals and symbols to gain power over the world through enlightenment. Secret organizations, like the Rosicrucian Order and the Freemasons, believed that an individual could progress through many stages of enlightenment to become almighty, almost like a god. The most magical art of all was remembering, which they believed was connected to imagination, to the divine creativity of humans.

When you think about it, it's not such a strange idea, because there really is something magical about our ability to store the past and retrieve it as lifelike images. Between our temples, most of us are equipped with our own private memory theater, which continually stages performances, always with slightly new interpretations—and now and then, with different actors. Today we know that everything we think and feel takes place in our brain cells, yet it is still almost impossible to grasp that our whole lives are to be found in our brains. So many emotions—fantastic, sad, beautiful, loving, and scary experiences—are hidden in our cerebral convolutions as electrical impulses, inaccessible to other people around us. Even people who have experienced the same thing have completely different memories of it.

BUT WHAT SORT of physical trace does a memory actually leave in our brain, and if we can locate it, can it explain memory? Memories are both abstract (states or episodes we can return to in our minds), and concrete (strengthened connections between neurons). Memories are incredibly complex. They are more than the trivia required to win a quiz show, more than the individual facts you look for among thousands of less relevant items in long-term memory. Just think of something you have experienced, recall your memory of it, and feel the sensations it contains. Are you watching it on your inner film screen? Do you hear the sounds, the voices; do you see the smiles, the eyes of the one you're talking to? Are you on the beach on a summer's day while the waves break against the sand? And the smells! Unlike at the movie theater, here we can smell the cinnamon buns and the ocean breeze, the seaweed in the bay, and hot dogs on the barbecue on the neighboring beach. You can even feel things, like the water hitting your body as you dive into the sea. All these sensations flutter about our brains as we remember. It's not possible to describe a memory by pointing to a few connections in the brain. It has to be felt.

At any rate, the hunt for the memory trace, the physical imprint of memory, has been a major part of brain research ever since neurons were discovered—well, actually, ever since Aristotle talked about wax seals. Some called it the *engram*, an inscription in the brain, and finding it became the holy grail of memory research. If we could find the engram, we would also understand the brain itself. With the help of our divers, we are trying to find the fishing net that holds our memories, the memory network. Every one of the squares in the net must be attached in some way; they are links that exist physically in the brain. Finding these links, and what they consist of, was a necessary step toward understanding how the brain handles memory. Before the 1960s, no one had succeeded in doing this.

A happy rabbit was perhaps all that was missing: Terje Lømo would find the very first memory trace, the smallest part of a memory, inside a rabbit brain. He is now professor emeritus in medicine at the University of Oslo and has worked mainly in physiology, the study of how the body works.

"I am most interested in how things work. Simply *describing* the brain was not enough for me," he says.

In 1966, he was leaning over a rabbit. It had once lived in the countryside, happily eating clover, without a care in the world. In the hands of Lømo, though, it now faced a problem. There it lay, sedated and with a fairly big hole in its brain, while the researcher came closer with tiny electrodes.

"We sedated the rabbits and sucked out a little of their cortex, so that the hippocampus was exposed. Then we poured warm, clear paraffin in the hole; it gives a good view, keeps everything in its place, and makes it warm and moist enough for the brain to continue working through the experiment. We had a window into the hippocampus."

His main goal was to find out what happened when he sent small electrical impulses through the brain, not because he was particularly interested in the hippocampus, but because that part of the brain was easier to observe. As opposed to the very complex cortex, the layout of the hippocampus was much simpler and more understandable, and the routes through it were already well known.

At the time, Lømo worked with Per Andersen, who had discovered that neurons could suddenly send off a train of signals, which were first measured by small electrodes used in experiments originally not concerned with memory at all. But neither Andersen nor other researchers knew what these signals meant. Now Lømo had decided to examine them more closely, which is where the happy—but soon dead—rabbit came into the picture. Lømo used a small electrode to set off tiny electrical impulses to travel from one part of the brain into the rabbit's hippocampus, where he measured the signals with a small receiver.

What young Lømo found was astounding and had never before been described. When he sent these electrical impulses through the rabbit's hippocampus in small "trains" of repeated signals, the cells at the other end eventually needed less stimulation to become triggered.

Some form of learning must have taken place; it was as if the neuron remembered that it was supposed to send its impulse when it had received the message from the preceding neuron! As if, initially, the first neuron had to nag it to send its signal: "Come on, come on, come on, fire already!" After having been prompted enough times, it understood to fire after just a cautious "Fire now!" And this response persisted. Something had permanently changed in the brain.

What he'd discovered was simply the smallest part of a memory, a tiny little memory trace. This response is now called *long-term potentiation*, meaning that a physical change occurs in some synapses in response to a recurring stimulus. At the same time as Lømo was making his discovery, neuroscientist Tim Bliss—a few thousand miles away from Oslo, at McGill University in Canada—had been looking for memory on a cellular level. What he lacked was the evidence that strengthened synapses were connected to memories. That is, until Lømo stumbled upon long-term potentiation! Bliss traveled to Oslo and the two did some experiments in 1968 and 1969, resulting in a scientific paper they published in 1973. Their paper presented a theory of how a memory is created on a micro level.

Almost nobody paid attention to the paper until twenty years later, because academia wasn't ready for it. There was simply no context; no other studies had trodden even close to this particular corner of research. Since then, though, Bliss and Lømo's paper has formed the basis for much of modern memory research. And now we know more: a memory consists of many of the connections they documented. One neuron can participate in many different memories. Memories are large networks of connections between neurons in the brain. When something becomes a memory, new links form—neurons either turn on or turn off, and either fire or don't fire a signal in the brain, and in that way form a pattern.

Our memories cannot all remain in the hippocampus, so they spread out across the cortex. It takes time before a memory matures and all the complex connections it requires to store all that makes a memory—smells, tastes, sounds, moods, and images—are established in the brain.

"Sleep is needed for a memory to consolidate. We believe that while we're asleep, we go through the events of the day in order for them to attach to the cortex. But when we are stressed, this doesn't always happen. The neurons don't fire in the same way. When I tried to re-create my experiment on other rabbits a couple of years later, it didn't work," Lømo recounts.

He'd been lucky the first time he experimented. His rabbit, despite its untimely end, had lived a happy life. The rabbits in the second experiment were stressed, so the neurons in their brains didn't work as they should have. In other words, you must treat your test animals nicely if you want to learn from them. The same goes for humans: when we are stressed, we don't retain memories as easily as when we are happy and relaxed.

At about the same time as Lømo's discovery, there were other breakthroughs in the hunt for the memory trace. In 1971, John O'Keefe at University College London found cells in the hippocampus that remember certain *locations*. For example, there are some cells in the hippocampus that are active only when we sit on a certain chair, and not on another chair—even in the same room. It is evidently up to some cells (*place cells*) to remember where we have been at all times. But to remember a place in and of itself—is that a memory? The Norwegian neuropsychologists May-Britt Moser and Edvard Moser—together with John O'Keefe— were awarded the Nobel Prize in Physiology or Medicine in 2014 for their work on that very question. The two Norwegians received the prize because they decided to develop

O'Keefe's research further and look beyond the hippocampus. Their work examined the entorhinal cortex, which connects the hippocampus and the rest of the brain. The Mosers experimented with rats, which, when they were free to explore their environment, showed cells firing in exactly that part of the brain.

With tiny metal electrodes surgically inserted into their brains, these rats wandered around their cages. A single neuron in the entorhinal cortex didn't react to just one place the rat had scurried to, like place cells, but to *several* places. Amazing, that what they were expecting to be place cells didn't remember only one location but several locations in the same area! But when the Mosers marked the points in the cage where the cells had fired, they formed a perfect hexagon on their computer screen. The more the rats ran around in their cages and mazes, the more obvious it became; on the Mosers' computer, a clear honeycomb pattern emerged. One cell, one hexagonal grid pattern. It was a coordinate system of the environment.

"At first, we thought there was something wrong with our equipment," Edvard Moser says. "The pattern that emerged was too perfect to come out of something real."

Each of these neurons makes its own grid, each slightly offset from that of the neighboring cells, so that all points in the environment are covered. Some grids are fine-meshed, while others react to points far away from each other, even farther than it is physically possible for the researchers to measure indoors. Without these *grid cells*, we are not able to understand or remember locations and where we are in relation to where we have been. We make these patterns wherever we go—wherever we stand, lie, or drive.

"We sent the rats into a ten-armed maze, and it turned out that they continued to make the grid pattern but also started a new one for each 'arm.' We believe that these patterns are patched together, so that the rats remember how to get through the maze," Edvard Moser says.

Since then, other researchers have found the same result in patients undergoing epilepsy surgery. It was as anticipated: in humans, as in rats, all locations are stored in a hexagonal pattern. We are all bees! We all organize the world around us as a hexagonal grid.

"We believe that this was developed very early in the evolution of mammals," Edvard Moser says. "And we believe that what we have discovered about grid cells is central for episodic memory. It is, after all, impossible to create memories without tying them to a place."

Other researchers agree that place and grid cells play a special role in episodic memory. Some go so far as to say that this system in the hippocampus and the entorhinal cortex has become specialized to assign each memory its unique memory trace, as part of a unique memory network. Perhaps, at first, the sense of place was the primary task for the hippocampus and the entorhinal cortex. But as evolution proceeded, our memory maps were given a new function: to take our individual experiences and tie them together in a grid. Hexagonal maps of the environment became hexagonal-patterned fishnets of memories.

Recently, researchers in California have been able to

demonstrate, in the hippocampi of mice, how memory networks link themselves to context-dependent memory. Like Terje Lømo, they made a window into the hippocampus, to the tiny little piece of it called *cornu ammonis 1*, Ammon's horn. Looking at the hippocampus in cross-section, it looks like a goat's horn, bent inward, into a spiral. Here, through the tiny window into the cradle of memory, the California researchers could see, under a slightly fancy microscope, how the neurons lit up when the mice were placed in different environments. They made three different cages, which would give rise to three different memories: a round cage, a triangular cage, and a square cage. The smell, texture, and other conditions also varied between the three cages. The crucial factor was how close in time the various experiences took place. Two groups of mice were compared. Half of the mice had a go at the triangular cage, and then directly afterward they were placed in the square cage. These mice got to experience two different cages in quick succession. The rest of the mice were placed in the round cage and then, seven days later, in the square cage. The second group of mice had two experiences—episodic memories—distinctly separated in time. When the researchers watched through the microscope while the mice were exploring the cages, they could see activity in the neurons in a defined area. Each of the three cages created a signature pattern of neuronal activity in the hippocampus, meaning distinct memories. The exciting part was that the experiences that took place close together in time led to activity in groups of neurons that overlapped. These two experiences hooked on to each other, not only in time but also in place, in the hippocampus of the mice. Meanwhile, when mice visited two cages a week apart, it was accompanied by activity in two separate groups of neurons in the hippocampus.

The researchers believe this happens because the activation of the one group of neurons causes other nearby neurons to become easier to activate. Everything links together in a network. The main point of Godden and Baddeley's context-dependent memory experiment has, in this way, been demonstrated in the brain—not in diving mice, but by diving into the cortex of the mice.

WHEN WE EXPERIENCE something—as we find ourselves in a specific situation at a specific location—and it becomes a memory in the brain, it spreads out across the cortex until we recall it. A memory is composed of thousands of connections between neurons; it is not one connection that makes a memory. A memory is more than Terje Lømo's long-term potentiation.

But what does a memory look like? Can we see a complex memory the way we can see a simple memory trace? To be able to do this, we must exit the rabbit and mouse brains and enter the human brain. And we must watch the brain while memories are recalled. Fortunately, we don't need to sedate humans and open their heads to get a glimpse of their memories. As we learned in chapter 1, Eleanor Maguire at University College London has used an MRI scanner and some reminiscing volunteers to observe the traces of their memories as they relive their past experiences.

An MRI machine uses a strong magnetic field to take pictures of the body. Different body tissues react differently to the magnetic field, which results in detailed images. The MRI machine can be programmed in a certain way to read the oxygen level in the blood flowing through the brain. Since neurons use oxygen to function, we can tell from the images where there's a lot of activity. We then know where in the brain nerve cells are most active while the test subjects remember things. This is called functional magnetic resonance imaging (fMRI)—images of the brain while it is working—as opposed to structural MRI, which shows us only what the brain looks like. The memories light up like tiny flashlights underwater, flashes that light up the sea in little spurts.

But is it really possible to see what memory a person is recalling? In Eleanor Maguire's laboratory, participants allowed themselves to be scanned by an MRI machine at the same time as they were asked to remember their own experiences. The professor actually managed to figure out what they remembered by studying the fMRI images. Maguire watched the activity in the hippocampus while the test subjects were thinking of episodes from their past, and she could see that each memory had a unique pattern of activity. She had a computer program that learned which of the test subjects' memories were tied to which patterns of activity. From that, the computer program could pick out which fMRI images hung together with which memories.

Is this simply a mind-reading machine?

"These are memories we had agreed with the volunteers,

before the scanning took place, that they would recall, not random memories. In a way it's, in very general terms, a kind of *voluntary* 'memory' reading," Eleanor Maguire says.

So far, she can see the track in the vinyl record, but she can't hear the music.

"The next step would be to be able to see what people remember without having decided on a fixed set of memories beforehand. But it's a long way until we get to that level," she assures us. We can safely leave mind reading to science fiction films and books.

Maguire isn't doing this because she thinks memories can be reduced to a checkerboard pattern in an MRI image. To her, memories are vastly complex—they are unique experiences that can only be fully known by the one who keeps them. They are also not static. She has observed that something happens to memory traces over time: two weeks after an initial memory is encoded, its memory trace is visible in the front of the hippocampus, but much older memories from ten years previous are processed further back in the hippocampus.

"Memories contain many pieces of the initial experience that are later brought back and put together again," she explains. "When the memories are still fresh, they are more easily accessible; we can easily picture the episode and how it happened. In the beginning, it is readily available within the hippocampus. As a memory ages, the pieces are stored in other parts of the brain and it takes more effort to reconstruct it and bring it back. The hippocampus puts all the pieces together in a coherent scene."

But what is she actually looking at? What gives the memories a unique "signature" in the fMRI images of the hippocampus? Eleanor Maguire believes that there are groups of neurons working together on one memory.

"The fact that we can see unique patterns for each memory must mean that information about the person's experience is present there; it has to be in some way related to the biological memory trace."

But because the resolution of an fMRI image is extremely coarse, we can only see large groups of nerve cells activated at the same time, as opposed to individual neurons.

"While it is important to study memory on a cellular level, we should also think of a memory rather like a big cloud of activity. A memory is more than the single synapses—it is much more complex than that," says Maguire.

To her, episodic memories are first and foremost about scenes. "All the little pieces that together make up a memory don't mean anything unless they are placed in a scene. The action *takes place* somewhere."

But as an episode is tied to a place and forms a scene in your mind's eye, an important component of this may be a set of grids—the map within the hippocampus and entorhinal cortex. The memory is tied down by all the little synapses being strengthened through long-term potentiation. Synapse by synapse, the memories are clicked into place.

"We are hoping that our discovery can help solve the enigma of Alzheimer's disease. Long before there are other symptoms, people with Alzheimer's experience spatial navigation problems," Edvard Moser says. The newest episodic memories also suffer first when the disease sets in. They go before all the knowledge we have gathered throughout life does, and also long before mature memories from long ago dissolve, like clouds of sparkling particles that swirl out to sea, never to return.

BUT WHAT ABOUT our divers? You haven't forgotten, have you, that we sent ten men down into the ice-cold water of the Oslo Fjord at the beginning of this chapter?

The rain is dripping from the eaves of the diving center here on dry land, and we're rubbing our cold and wet hands together in a futile attempt to stay warm, our teeth chattering. The divers are, of course, voluntary participants; nobody is forcing them to do this. Still, with only a few remaining bubbles on the surface reminding us that they are down there, it's easy to be a bit worried. What if something were to happen? And what if they remember as poorly as, well, a jellyfish? We will return to the divers shortly, but since we brought it up: Do jellyfish remember?

"We don't know if jellyfish remember," biologist Dag O. Hessen says. "But jellyfish do have a kind of 'will,' since they swim in a certain direction, even if they don't have a brain, only nerve fibers. However, all animals, even the simplest ones, have a certain capability to learn."

How did human memory become as advanced as it is? Why do we remember the way we do and not the way jellyfish do? What might the alternatives have been?

"We have not been able to prove that animals have memories that work like human memories. We believe that other animals' memories associate to a situation and pop up when they see or feel something, as when for example a cat sees a cupboard door and remembers that it hurt its tail there once," Hessen explains.

So there's no proof that zebras can stare melancholically into the sunset and remember the great loves of their lives, or that a dog can suddenly bark mournfully because it's thinking of a sad episode from its youth. No gazelles cringe because they're thinking about an embarrassing moment two years ago, no leopards experience a flash of happiness when a memory hits them of how they killed their first prey. At least, not that we have been able to prove.

"We believe that only humans do this: look back in time regardless of context. All animals and plants have some form of memory, in the sense that they adapt to the environment. It's beneficial to learn to avoid dangers and remember how to secure food and partners. It's obviously an evolutionary advantage for all living organisms, even for short-lived ones, to be able to remember and not only live in the moment. What's special about humans is our ability both to see the past before us and to create visions of the future. To be able to envision the future is possibly a byproduct of memory," Hessen believes.

The biologist suggests that there's another reason for humans to have developed a large brain with advanced memory, something that has to do with our social groups.

"We know that social animals have larger brains and more memory than animals that aren't social animals."

An example: all bats are, in a sense, social animals, but vampire bats are particularly social. They live in groups, and they can't survive for more than three days without fresh blood. According to Hessen, researchers have found that bats—sympathetically enough—help each other by regurgitating blood for others, even bats outside their family, and it seems as if they remember favors that have been done for them earlier. There's a form of reciprocity between vampire bats that's very similar to humans, like friendship.

"Many believe that humans have good memory because people are social animals with many hierarchies and exchanges of favors. Sympathies and antipathies depend on remembering. And the longer one lives, the more one has to remember complicated social structures," says Hessen.

Animals that live longer remember more. An example is the elephant. It does actually remember—like an elephant.

This is just one of many anecdotes about elephant memory. In 1999, as the zookeepers in the elephant sanctuary at Hohenwald, Tennessee, introduced their elephant Jenny to a newcomer named Shirley, Jenny became agitated. Shirley also seemed more than usually preoccupied with Jenny. The two elephants behaved as if they knew each other. Upon investigation it turned out that, for a short while more than twenty years earlier, the two elephants had worked together in the touring Carson & Barnes Circus. According to Hessen, researchers that have followed elephants over long periods of time have found that elephant herds are highly dependent on good memory. The matriarch of the herd must be old enough to have the experience to lead her

herd to safety if there is a fire or to find water during dry
spells; younger matriarchs risk making fatal mistakes.

The elephants Shirley and Jenny acted as if they really
had *human-like* emotional memories of each other. But
memories can also be far less complicated, without being
less impressive. Several animals have a kind of instinct—
or memory—for time and place. Puffins return to the west
coast of Norway on exactly the same date year after year,
regardless of weather. American and European eels swim
all the way to the Sargasso Sea to spawn. Monarch butter-
flies have multiple generations each year, of which only one
lives long enough to migrate south and back. It's impossible
for the new generation of migrating monarchs to remember
where their great-great-grandparents came from, but still
they know to go south to particular wintering grounds. Is
this memory or instinct? And can instinct be tied to a certain geographic location or a certain date?

"When the salmon returns to the spawning grounds it
came from, it uses its sense of smell, and the sense of smell
is closely related to memory in most animals. But there's
a lot about animals' memory that's still a mystery to us, as
for example this thing about the eel," Dag O. Hessen says.

Even in the human brain, the so-called olfactory bulb
is situated close to the hippocampus, pointing to the fact
that smell is the sense most closely tied to memory. This
doesn't mean that the other senses aren't strong too. Marcel Proust came up with his seven-volume work when he
tasted a madeleine cookie dipped in tea. For many, sounds
and music are tied to strong memories; just think of how

an advertising jingle can stick in your memory. How many thousands of tunes are we familiar with?

Songbirds are birds with good memory. Just like us, they have to learn tunes—they aren't born with them. A songbird placed in some other songbird's nest will learn the wrong tune; a blue tit that is placed in the nest of great tits will learn the great tit's tune. The songs of songbirds can have both dialects and other variants. The European pied flycatcher, for example, varies its tune according to its intended recipient: the "wife" or a "mistress." What makes a bird's memory especially impressive is its brain. Birds have several song centers in their brains, one of which is the higher vocal control center, which grows each spring and has almost completely receded again by the fall!

"We don't know why it happens, because the birds remember the tune they've learned even without the higher vocal control center," ornithologist Helene Lampe at the University of Oslo tells us. There is still a lot scientists don't know about this avian brain region. Female birds typically don't have particularly well-developed higher vocal control centers yet are still able to sing. It's believed they have it so they can identify and remember rivals, but in the case of the European pied flycatcher, it does the female no good; she watches the nest while the male is out looking for more "mistresses."

"This is a songbird mystery we still haven't solved. We don't know where the song is actually stored, but recent research points to the auditory center of the brain being used for some storage," Lampe says.

Many types of birds remember amazingly well: migratory birds remember where to go, parrots and crows can learn human language, and jays that cache food find their way back to their stashed nuts.

"Hoarding requires a good episodic memory—that is, having a vivid memory of the act of burying the nuts. Remembering this experience makes it possible to find them later," Lampe says. Herein lies one of the great controversies in memory research: How uniquely human actually is episodic memory, and can we find evidence that other animals and birds also have this form of memory? Scientists don't have a final answer.

We take our way of remembering for granted. The human, or mammalian, way of connecting experiences through long-term potentiation, creating large memory networks that are kept in place by the hippocampus, could be just one way of doing it. Nature has a wealth of alternatives to offer. Animals without hippocampi also have memory. Even one-celled animals, like slime molds (Mycetozoa), show signs of remembering. In one experiment, researchers exposed a slime mold to moisture and drought on a regular basis and watched it react. After a while, they stopped stimulating it this way, but it kept reacting at the same intervals as before, for quite some time. Slime molds have even found the quickest way through a simple maze! Amoebas leave slime where they've been, so that they don't reenter a dead end in the maze but rather explore new paths. They wander through the maze with their one-celled memory, never knowing that evolution has raced past them.

Slime molds, jellyfish, songbirds, eels, monarch butter-flies, vampire bats, puffins, and elephants represent different mysteries when it comes to memory. Which ones have memory, and which just have instinct? They each show us that there are many ways nature can meet the need to keep information for later use. But human memory is perhaps the greatest and most complex. What other animals remember episodes from not only their own lives but also their ances-tors' lives from many thousands of generations ago, and record their memories for others to read and remember?

THERE ARE ENOUGH mysteries within our own mem-ories to keep us busy. Take, for example, Henry Molaison, who opened up so much research into memory: How could the man without hippocampi remember his life before the surgery? As we know, memories appear in the hippocam-pus when we retrieve them; they light up on the screen of Eleanor Maguire's MRI machine, where they create differ-ent patterns. How was it possible for Henry to remember anything at all without his hippocampus, when it is the hip-pocampus that reassembles memories? This is something memory researchers are still fighting about. The fight is as big as the battle about the role of the hippocampus in memory.

Henry's memories prior to the surgery had gone into storage the normal way, with the help of the hippocam-pus. His memories had consolidated as memory traces tied experiences together. Later, the synapses in his cortex were strengthened, until they could manage without the help of the hippocampus. This process may take many years. That's why Henry didn't remember anything from the last couple of years prior to the surgery. The memories from this period were simply too unstable and dependent on the hippocampus. For a long time, it was believed that this was the full explanation and that the hippocampus wasn't necessary at all when it came to recalling early memories. But then researchers, like Eleanor Maguire and others, started to notice that things happen in the hippocampus when we retrieve a memory.

They didn't question whether or not Henry's memories were real, but they did point out that a memory is not just a memory. A memory may have turned into a story that includes facts about what happened, not unlike an anecdote. On the other hand, a memory can also be something completely different: a re-creation of the experience, filled with sensory experiences, emotions, and details of how the episode unfolded in time and space. Henry's memories were probably more like the first kind, resembling book knowledge or simple tales, called semantic memory. He seldom gave particularly detailed descriptions of his childhood. Often, the stories began with "I used to… ," followed by facts about where he'd gone to school, where he'd vacationed, and who his family was. He possessed a rather dry encyclopedia about himself. Presumably, he could not recall lifelike, smelly, noisy, emotional memories. After having known Henry for years, researcher Suzanne Corkin was convinced that his memories lacked the vividness so characteristic of episodic memories.

BACK AT GYLTE Diving Center, we've split the divers into two groups and numbered them from one to ten. The divers are completing their first memory test, the one we use for comparison, measuring their normal memory. The men are visibly sweating over the twenty-five words we gave them to remember. Not only because the test is hard; they have to look at their list of words for two minutes, then go for a little walk and return to the table to write down what they remember. But with the diving gear already halfway on, they are hot and perspiring more than they might like. The divers manage to remember between six and seventeen correct words, completely normal results.

That day by the fjord, the rain on our skin feels like pins and needles of nervousness as the first group goes down into the water. What if we don't find out anything at all? What if the men are diving in vain and don't get to prove anything about memory and context?

Of course, we can't go through life relying on our surroundings to help us remember everything. Godden and Baddeley also pointed out that this was an unreasonable idea. In *An Essay Concerning Human Understanding* (originally published in 1689), the philosopher John Locke described a man learning to dance in a room with a large trunk. He could do the most elegant dance steps, but only as long as the trunk was there. If he was in a room without the trunk, he was hopeless on the dance floor. This sounds very strange, and fortunately the story probably isn't true. It highlights, though, the idea of context-dependent memory. The point Godden and Baddeley made was that our memory may rely on context *to a certain degree.* Can this be useful in some way? Should we cram for an exam in the location where we'll be taking it? Or remain in the same apartment until our dying day for fear of losing the memories that have been made there?

Fortunately, we do have access to our memories when we are not in the same environment as where we experienced an event. The divers at Gylte can recount the amazing experiences they've had in the water even when they are safely on shore.

Our memory networks—our fishnets of memories— benefit from context beyond just our physical surroundings. We create the strongest memory networks on our own, when we learn something truly meaningful and make an effort to understand it. Someone who is passionate about a particular subject, such as diving, will more easily learn new things about diving than about something she's never been interested in before. This is because she already has a large memory network devoted to diving where she can store her new knowledge, and because she is motivated. It's as if we can add another layer of netting just because the self is involved; memory is self-serving. Memories are linked to what concerns you, what you feel, what you *want.* Too bad, then, that so much of what we actually need to remember is so darned uninteresting!

Lately, others have tried to test context-dependent memory in other ways. Do we remember things we learn while skydiving? The researchers concluded that the stress level of skydivers was so high that it erased all effects of context. This may not be so strange—if we are so high on adrenaline that we barely notice where we are, there are no surroundings to support those memories. More practical were the researchers who wanted to examine if medical students remembered more when they were in the classroom where they had first been taught. The classroom, in this case, was either an ordinary classroom or an operating theater, where the students were dressed for surgery. Fortunately for the future patients of those medical students, it turned out in the experiment that the differences were so minimal that doctors can safely continue practicing medicine far from the context of learning.

In our experiment at Gylte, we split the divers into two groups. The divers in the first group would be tested on what they remembered on land after trying to memorize twenty-five words underwater. The others had to both learn and recall the words underwater.

The five divers in the first group come ashore, splashing as they go. They wriggle out of their masks and flippers, unhook leaden oxygen tanks and sit, legs spread apart, on the bench along the wall of the Diving Center.

Their results are miserable.

One of them remembered only words from the first test—the one for comparison—and got a zero on the underwater words. The best one remembered thirteen words from the list he saw underwater, but this too was worse than he'd done during the first test on land. The average result of the comparison test, inside the Diving Center, was 8.6 correct words. The divers remembered an average of 4.4 words when they emerged from the water.

"I sort of thought I had the words there while I was underwater, but then we got up on land, and it was as if my mind changed completely, and I lost it," one of the divers says.

The removing of flippers, tottering up from the edge of the pier along the walk to the Diving Center, lifting the tanks from their backs, and grabbing a piece of paper may of course have disturbed their trains of thought and pushed the words out of the way. Duncan Godden and Alan Baddeley had pondered this possibility and tested whether all the trouble of getting back onto dry land could have thrown off the result. They let one group of divers learn the words on land, then dive and come up again, and compared them with a group who'd learned the words on land and waited the same amount of time, but without moving. The group that dived in the middle remembered just as much as those who had remained still in the same place. So all the hassle of changing location could not have explained why the divers who learned in the water remembered less on land.

Deep beneath the surface, the divers in the second group have taken out their flashlights and waterproof notepads that make it possible for them to write underwater. Bubbles from their breathing pop on the water's surface; they are fifty or fifty-five feet down, and it's hard to handle the plastic-covered sheet to write the twenty-five new words. They've gathered in a circle in the dark, and short flashes from the flashlights tied to their arms shine through the water every time they move their hands and write. Like the words they learned on land, these are mainly one-syllable words: short and concrete and easy to write with gloves on.

This group had remembered an average of 9.2 words when they were tested in the Diving Center. But what happened when they tried to learn twenty-five words underwater and were supposed to remember them underwater? As the bubbles grow bigger and the divers slowly rise to the surface, those of us on the pier are long since soaked through and clinging to empty and wet paper coffee cups. Even the seagulls have stayed at home today.

The divers, on the other hand, are not in a hurry. They rest for a while a few feet under the surface before they get out of the water. Around us, clumps of old snow lie between tufts of rotten grass. Our excitement has been building this whole ice-cold morning, as has our longing for hot chocolate and dry socks; the divers, however, are satisfied with their dive. They proudly hand us their notes.

When we examine the results, it occurs to us that we have managed to re-create the experiment from the 1970s almost down to the smallest detail. The divers who were supposed to learn *and* remember underwater have remembered on average 8.4 words in the deep, almost matching their achievement on land earlier that day. They pulled this off despite factors like increased pressure underwater, gas mixtures and masks and wet suits and the sound of breathing, clouds of bubbles swirling toward the surface, flashes from flashlights sweeping the bottom of the sea, blurry vision, uncomfortable wet gloves, and difficulties holding pens and waterproof notepads. In the famous experiment from the 1970s, it was clear that the context had an obvious effect—the divers had remembered the list of words much better in the water when they had also memorized it in the water. Actually, they remembered it equally as well as the list they memorized and recalled on land.

When the divers were in the water, they recognized where they had been before, and this memory triggered the memories of what they had learned, so that the words popped up almost by themselves, like images on a screen.

Caterina Cattaneo led the divers in our experiment. She has almost thirty years of underwater experience and has dived at a depth of two hundred feet. This was a simple dive for her. The water temperature was comfortable, she claims, as she swings herself up on the pier and wrestles herself out of her diving mask. The February rain sprinkles the fjord behind her.

"I've never seen seahorses here," she tells us. "I've seen two on Madeira. They were tiny and very cute. They bobbed up and down, their tails wound around a sea plant. But the current was strong, and suddenly I was far away from them. I only caught a glimpse of them."

THE SKYDIVER'S FINAL THOUGHTS

Or: What are personal memories?

_____

All the flowers in our garden and in M. Swann's park,
and the water-lilies on the Vivonne and the good folk of the
village and their little dwellings and the parish church
and the whole of Combray and of its surroundings, taking
their proper shapes and growing solid, sprang into
being, town and gardens alike, from my cup of tea.
MARCEL PROUST, *In Search of Lost Time*

FOR MANY YEARS, our sister was an active skydiver.
Every weekend she would set out for the skydiving field
at Jarlsberg or travel to the United States or Poland to
jump in large formations with hundreds of other skydivers.

Watching Tonje skydive was often a terrible experience
for us. During the minutes we observed her falling from the

sky we imagined her funeral, complete with flowers and the music we'd choose to play as the coffin was carried out. Even though there are few accidents in skydiving, the few that happen are gruesome. You don't plummet toward the ground from fifteen thousand feet without it being dangerous. Every time she was about to land, we drew the deep sigh of relief that comes after holding your breath for too long. The cheeriness of the large, brightly colored parachute belies the grim reality of the accident it can cause if it doesn't unfold or if a sudden gust of wind grabs hold of the lightweight material. Tonje's parachute was reddish orange, like a sunset.

The plane drones so loudly on its way to jumping height that you have to shout to be heard. This Saturday, a day in July 2006, Tonje walks toward the open door of the little silver plane, a Soviet turbine machine, Antonov An-28. She positions herself on the edge. She believes all will go well; she can't possibly think anything else or she wouldn't throw herself out of a plane thousands of feet above the ground. Normally, it does end well; that's the thought you cling to.

Now we'll leave her standing there, watching the forested, billowing landscape from above, while thick clouds cast everything below her in a grayish light. The temperature hovers around sixty degrees Fahrenheit, and the summer has not yet fully set in. We'll let her stand there for a while longer, her slender body in her red skydiving suit, her dark brown eyes, her broad smile. Just a few minutes more.

What memories would you linger on if you had only a few moments left to live and were looking back on your life?

What memories are like shiny pearls in an incredibly exclu-sive—exclusive, because you are the only one in the world with your memories—pearl necklace of important events? What flutters across the hippocampus as you say goodbye to life? How many monarch butterflies light on your hand?

Or what if you were allowed to pick only one, as in the Japanese film *After Life*, where the deceased have to choose a single memory to relive, over and over, in heaven—the happiest moment of their lives. What would yours be?

Perhaps this is why people keep diaries. They don't want the magical moments to slip away.

When blogger Ida Jackson reads through what she has written, she remembers more of those days than before, she claims. She sees and smells and hears what happened. She discovers details she otherwise would not have remem-bered. She is, in a sense, a collector of memories, a memory hoarder.

"It feels as if, by doing this, I lose fewer memories. There is something existential about it. I think often about death, so I want to remember everything," Ida says. From 2007 to 2010 she wrote the award-winning blog *Revolusjonært roteloft* under the pen name Virrvarr; it was Norway's third-most-visited blog. She saw it as an extension of her diary. She has kept a diary every single day since Christmas of 1999.

"Today, I got this notebook in the mail, and since my life is upside down right now, I might as well leave behind something in writing," is how twelve-year-old Ida Jackson began her first diary. Since then, she has written herself into the long tradition of diarists and autobiographers, philosophers, poets, and authors—from St. Augustine to Karl Ove Knausgård—who have transformed their lives into books, published or not. It seems as if written language is closely connected to our wish to remember. The first Babylonian writings from over four thousand years ago were memos, trade notes, and astronomical calculations etched into ceramic plates meant to be kept for posterity.

By the year 200 CE, emperor and philosopher Marcus Aurelius had written what is considered the earliest well-known diary, *Meditations*. But long before that, Japanese courtesans and other Asian travelers were in the habit of recording their experiences in writing.

So what do we remember of our lives when we write things down, or when we don't?

Psychology professor Dorthe Berntsen heads the Center on Autobiographical Memory Research in Aarhus, Denmark, and exclusively researches personal memories. "We remember best the period from our early teenage years into our twenties," she tells us.

It seems that not all memories are created equal. Some are given priority. Our memories peak during our formative years (teens and early twenties), a phenomenon called the *reminiscence bump*. During this period of our lives, many of our experiences are new and startling; there are so many firsts, and they stay with us for the rest of our lives. Middle-aged people who are asked to recall their fondest memories typically mention something from this period of their lives, Berntsen's research reveals. This area of psychological research is amazingly free from controversy.

But does it help if you keep a diary the way Ida Jackson does?

"Yes, it does help, but it might mean that we replace our memories with written stories," Berntsen says.

What else helps make memories last? It turns out that several factors determine whether an experience sticks with us as a memory.

One such thing is the emotional impact of an experience. Exciting events that provoke sharp emotional highs stay with us particularly well. Whizzing toward the earth from fifteen thousand feet in the air, for example. Or a first kiss, after anticipation has been building for weeks. Another important ingredient for a lasting memory is how much it deviates from what we expect—how distinct or remarkable it is.

Many memories are similar to a number of other experiences we have had. It's hard to tell them apart—thinking of them doesn't remind us of one specific incident. Like all the times we take the bus to work. We have a cumulative memory of these experiences under the heading "bus trip to work." Or all the times at the beach that have merged into "sunbathing at the beach": that feeling of a summer breeze grazing our face while we squint at the sun. This is not a single event; it has happened many times. Every time, we have soaked up the summer warmth and wished that the moment would last forever. Caterina Cattaneo added to her memories of diving when she dived for the seventy-third time: that feeling of sinking into the dark water, the bubbles rising toward the light surface, the maneuvers with the tank she had done seventy-two times before. All of this became part of the general memory of "diving," "diving in the Oslo Fjord," or "winter diving." But more exciting events remain as independent, unique memories. Like the time Caterina saw a rare marine slug for the first time, or the time she saw a seahorse in Madeira.

"The brain works with memory on two conflicting principles," psychologist Anders Fjell, from the University of Oslo, points out. "Part of the brain's work is to try to categorize and assimilate as many of our experiences as possible in order to save space, while the hippocampus fights to retain unique memories."

The hippocampus is finely tuned to notice and pick up events and experiences that stand out for being different. Their uniqueness is what creates a memory trace, a shiny pearl in the necklace.

As with all other information we encounter, the more we ruminate over and talk about a unique incident, the more ingrained it becomes in memory. All the little tales about our lives that we share around the lunch table, at parties, or on Facebook—small talk—make memories stick. The paradox is that those memories then become stories in our minds more than living experiences.

Dorthe Berntsen's research center is situated in Aarhus, Denmark. There, on top of the AROS Aarhus Art Museum, you can enjoy a view unlike anything else in the world. On the roof, artist Olafur Eliasson has installed a circular tunnel of glass in every color of the rainbow. In every direction you look, you can see spires and low-lying stone domes that date back to the 1600s in various shades of red, orange, yellow, green, blue, indigo, and purple, depending on where on the roof you stand. Just as the city of Aarhus is rendered here in multiple beautiful shades, our memories are also seen through a filter: our emotions.

The fate of a memory is mostly determined by how much it means to us. Personal memories are important to us. They are tied to our hopes, our values, and our identities. Memories that contribute meaningfully to our personal autobiography prevail in our minds.

Personality and identity can also be maintained without memories. Even Henry Molaison, the man without a memory, obviously had a sense of a *self*. He knew who he was, even though he didn't remember the full story of how he had become that person. Who we are is partially determined by factors like temperament and habits, and how we face the world and all its challenges. But our core memories in our own personal autobiography define us. Even if we don't write six volumes about ourselves, as Karl Ove Knausgård has done, all of us walk around with an autobiography stored in our memory. It isn't just a random stream of events we have experienced; our memories are structured and organized in accordance with our own life story. We are all authors.

"A *life script* is what we call it in memory research," Dorthe Berntsen says. "It's a script for how life should unfold; it structures our experiences."

If you ask children what they want to be when they grow up, they might answer police officer, firefighter, or doctor, or maybe author, psychologist, or skydiver. In other words, they know that adult life includes a job, perhaps even marriage and children. Before we even start school, we understand that life has direction. Our life script contains expectations of how life normally looks, with milestones such as starting school, getting a driver's license, graduating, starting a career, getting married, becoming a parent, and retiring. Gradually, as life progresses and we adjust our expectations, our life script also helps us access our memories by providing chapters we can browse in our book of life: "School," "Marriage," "Work," "Skydiving." When we activate part of the life script, we activate all the related concepts in the network it's part of, just like marine slugs, air bubbles, flippers, and seaweed triggered undersea memories for the divers in our diving experiment. When something reminds us of our student days, we mentally travel back to the student cafeteria, making it possible for us to remember many experiences from that time, especially emotionally loaded memories—ones that stood out, that we thought about often and discussed.

"We can't walk around remembering everything we've done in life all the time," Dorthe Berntsen emphasizes. The life script gives us an overview of life. It portions out our memories. Should we search in the wrong chapter, we won't find the memories we are looking for. Parts of our life history are therefore not always available to us right in the moment. When we enter a new chapter of life, it takes more effort to retrieve memories from an earlier chapter.

Stepping outside our life script comes at a cost, something astronaut Buzz Aldrin knows well. He was the second human in history to set foot on the Moon, an event that turned his life upside down. His personal memories are, to say the least, remarkable. There are not many who can look up at the Moon and reminisce! In one of his memoirs, he describes his lunar memories in vivid detail:

"In every direction I could see detailed characteristics of the gray ash-colored lunar scenery, pocked with thousands of little craters and with every variety and shape of rock. I saw the horizon curving a mile and a half away. With no atmosphere, there was no haze on the moon. It was crystal clear."

As Buzz Aldrin is about to set foot on the Moon, he takes his time to absorb some of the impressions the beautiful view offers: "I slowly allowed my eyes to drink in the unusual majesty of the moon. In its starkness and monochromatic hues, it was indeed beautiful. But it was a different sort of beauty than I had ever before seen. *Magnificent*, I thought, then said, 'Magnificent desolation.'" This description became the title of one of his books about the Moon landing, *Magnificent Desolation: The Long Journey Home from the Moon*, from 2009.

When Aldrin began training as an astronaut, he had his sights set on the Moon, and everything he did became part of a new life script that included landing on the Moon. The life script contained earlier chapters from his time serving in the air force and studying to become an engineer, the natural introductory chapters to his personal saga. But representing NASA—and perhaps becoming a large part of the United States' Cold War identity—was not part of the original script. Aldrin dealt with the strain of being in the spotlight by consuming alcohol. In description as detailed as his memory of the Moon landing, he relates his memories of his first glasses of whiskey and the feeling of calm they brought him. Sinking into alcoholism was far less heroic than traveling to the Moon, but fighting his way out of it was equally brave. And for that, there was no script.

"How did it feel to be on the Moon?"

Buzz Aldrin has been asked that question thousands of times. It's the world's best opening line, one would think. To Aldrin it has become as familiar as a broken record, and he won't answer the question any longer.

"I have wanted NASA to fly a poet, a singer, or a journalist into space—someone who could capture the emotions of the experience and share them with the world," he writes. Still, it would be incredibly interesting to find out how his memories from the Moon have affected him through the years. Are they memories he consciously retrieves and enjoys? Does he reexperience the excitement he felt right before the *Eagle* touched down on the surface of the Moon? Do memories from the Moon appear spontaneously in his daily life? Does he walk on the Moon in his dreams?

Psychology professor Dorthe Berntsen examines, among other things, spontaneous memories in her research. These are memories that appear on their own, without our consciously searching for them. But how do we capture a person's personal memories in the moment? Berntsen is interested in the average memories of ordinary people who haven't performed extraordinary feats, in outer space or elsewhere. To research spontaneous memories, she gives her subjects a timer and a notepad to carry around with them as they're going about their normal day-to-day activities. When an alarm sounds, she asks them to write down whatever memory comes to mind. She found that what people often remember is something their environment reminded them of. Spontaneous memories are not unlike a cat's memory when it sees the cupboard door that once closed on its tail—and jumps. For people, though, the associations are much more complex. The environment is full of potential cues that may trigger obscure memories. The things we see, and also smell, taste, talk about, and hear—particularly music—are paths into memory.

"Remarkably often, music is mentioned as a trigger for a personal memory," Berntsen tells us.

When her test subjects share the memories they had during the course of the day and when they had them, they point to music on the radio as a typical cue to a particular memory.

Play the music you loved listening to when you were young, and see if you are not suddenly back in the place where you first heard it. The feeling and the mood can come on so strongly that you suddenly remember smells and colors, clothes and details from your home, things you thought you had forgotten.

"Soft music began to flow from the ceiling speakers: a sweet orchestral cover version of the Beatles' 'Norwegian

Wood.' The melody never failed to send a shudder through me, but this time it hit me harder than ever," is how Haruki Murakami's *Norwegian Wood* begins.

The book is a nostalgic love story woven through with symbolic meaning from the Beatles song by the same name, and the opening of the book describes the strong memories music can evoke in us. Whole landscapes and stories can appear, unbidden, in our awareness.

It is well documented that music, which speaks so directly to our feelings, is a powerful memory cue. But what about smell? The olfactory bulb, which allows us to perceive odors, is located very close to the hippocampus. We may forget it sometimes, but humans are animals, and animals depend on their sense of smell to avoid danger. Why, then, isn't smell the best key to our personal memories? But smell *is* an important cue. Berntsen's research shows that our sense of smell is particularly important early in life. Perhaps this has something to do with childhood memories being less tied to our later interpretations and stories about ourselves, allowing more room in our memories for smell, which is more immediate and sensuous. Or perhaps it is because the odors we smelled in childhood aren't ones we encounter every day. When we get a whiff of childhood, it's a potent trigger for a distinct memory trace, because it hasn't been watered down daily during the years that have passed since we last smelled it. It is a time capsule which takes only a moment to send us back in time. Think about this: Can you remember the smells of your childhood home?

In Marcel Proust's *In Search of Lost Time*, he opens up a world of memories when he soaks a madeleine cookie in weak tea. Taste and smell are similar gates to the land of childhood.

"Apparently, Marcel Proust's trip into memory did not start with him eating a madeleine—a disappointingly tasteless cookie; tasty, but not distinct. Proust was eating toast, but along the way, he replaced toast with a madeleine cookie. A piece of art is more than just memories; it gives the memories a form," says Linn Ullmann, who in her novel *Unquiet* explores her childhood memories and her relationship with her father, the world-renowned director Ingmar Bergman.

The path into her memories followed the winding road of free association, not the logical archival approach one might have chosen when writing an authorized biography, yet her method is the one that best mirrors the way memory works. A life history can just as easily unfold while chasing a white rabbit as by following the historian's strict logic. Ullmann's research period was thus not spent scouring the comprehensive archive of her father's letters and documents, but by following her emotions and immersing herself in art and music and dance, putting herself in the right mood for the book she was about to write.

"Writing about memories is hard work; it's more than just transcribing recollections. I used to think I couldn't remember anything in particular from my childhood, but when I began writing I could conjure up complete episodes," she tells us. The memories Ullmann describes in her book are malleable. They are not static archives that hold perfect representations of things she has experienced. Because memories take many forms, we can approach them in different ways.

"Like the choreographer Merce Cunningham, I am thinking about what happens as our eyes follow the motion of a body from center stage to the outer edge. When I write, a small motion can suddenly become important, and something larger can become insignificant," she says.

In her book she describes how she celebrated Christmas the only time she ever spent it with her father. She is newly divorced, he is a recent widower. They walk through the snow from his small apartment to the Hedvig Eleonora Church in Stockholm. The snow whirls in front of their faces and around the church spire. She describes how, for a long time, she thought that he needed her because he didn't want to spend Christmas alone. With time, her understanding of that night changed. He always celebrated Christmas alone—in fact, he preferred it that way. It was she who needed him. The memory turns itself around and becomes another memory.

"I can't remember if the snow really fell that way. At the end of James Joyce's 'The Dead,' he describes the kind of snowy weather I am talking about. I don't know if it's actually his snowy weather I wrote into my story. However, it doesn't matter; things I have read and things I have experienced have blended together; I am not writing a biographically true story," she tells us.

Why do so many authors draw on their own memories? Maybe there is something authors can teach us about memory?

"Memory is a basic survival tool. We use it to tell stories about who we are; we *are* our own stories. Our love stories help us build our romantic relationships. On birthdays and anniversaries, people make speeches about things we have done. We tell stories about ourselves and about each other, on a personal level, on a national level, and an international level, as cultural stories. But our memories are actually fragmented, special, and creative! Memory is a force that both creates and preserves, because it writes new stories at the same time as it maintains our lives in little time capsules. For me, as an author, it is an exciting and unreliable tool. I often remember incorrectly," she says.

What Ullmann does is not unlike what all of us do, all of the time: we make things up, structure and transform, and suddenly our memories include things we haven't really experienced—just read, seen, or heard. Like James Joyce's description of snow that wove itself into the tale of the walk to Hedvig Eleonora Church. Memories *are* unreliable.

"I wanted to see what would happen if I allowed us to emerge in a book as though we didn't belong anywhere else. For me it was like this: I remembered nothing, but then I came across a photograph of Georgia O'Keeffe that reminded me of my father. I began to remember. I wrote: 'I remember,' and felt unnerved by how much I had forgotten. I have some letters, some photographs, some scattered scraps of paper, but I can't say why I kept precisely those scraps rather than others, I have six recorded conversations with my father, but by the time we did the interviews he was so old that he had forgotten most of his own and our shared history. I remember what happened, I *think* I remember what happened, but some things I have probably made up, I recall stories that were told over and over again and stories that were told only once, sometimes I listened, other times I listened with only half an ear, I lay out all the pieces next to each other, lay them on top of each other, let them bump up against each other, trying to find a direction," Linn Ullmann writes in *Unquiet*, almost as a report on how she has used memory as a method.

The debate about what autobiographical fiction really is, compared to autobiography, has been going on for a long time—long before Karl Ove Knausgård wrote *My Struggle*. But the raw material in both cases consists of memories. In autobiographical fiction, memory triumphs over hard sources and personal experience has greater value than objective fact. Memory, with all its creative misinterpretation, gets top priority.

"I discovered that memory isn't a locked trunk full of true recollections, but a creative sponge—it absorbs everything around it and renews itself," says Ullmann.

Ullmann's book is an exploration of the conundrum of constructive memory. What is actually true of what she remembers? While her father was alive, he talked to her about Bach's cello suites and described the saraband, one of the movements, as being like a painful dance between two people. Ullmann's book was inspired by their conversations about Bach. The book has six parts, just like the six parts of Bach's fifth cello suite.

In her book, Ullmann writes: "*To remember* is to look around, again and again, equally astonished every time." She probably wasn't aware of how right she was from a scientific point of view. Our personal memories are always reinventing themselves; new details are added all the time. She says that turning her memories into a novel involved both artistry and hard work. "Nothing is more boring than listening to someone who has just woken up tell you about a dream. It only means something to the one telling it. A dream can be an interesting experience, but it's not art. It is the structure that makes it into something more," she says. Memories, too, need conscious elaboration to become literature. What she thought of as a fragment of a memory might have become several pages as she reconstructed it factually and artistically. The title of *Unquiet* may very well allude to the fundamental nature of memory. Memories are not static, not authoritative, not solid as mountains. They are diffuse, they move around, they collide; they are like seahorses dancing restlessly amid the seagrass. Memory is constructive; it picks up fragments of an experience and builds a framework, a story about what happened. Once, that experience was fresh in our minds. But our senses, our attention, our ability to interpret, and our memory did not manage to absorb everything down to the smallest detail. Still, when that memory is retrieved, it seems as if it is intact. The memory itself becomes a new moment in consciousness, although as from a parallel reality. But beyond our perception, both hard work and artistic effort lie behind each memory.

The recorded conversations between Ullmann and Ingmar Bergman offer an illusion of truth. They represent exactly what was said, etched into a digital memory. But are they actually more truthful than the thoughts, feelings, and experiences that also form part of the memories she describes in the book? Are your Instagram and Facebook accounts telling the truth about your life, or is the treasure trove of unreliable memories you carry within your temporal lobe?

The early psychologist we met in chapter 1, William James, was something of a celebrity in late nineteenth-century America. He thought memory was the relationship between what we see and hear (we'll call that A), and what we store and can retrieve (we'll call that B). The way he explained memory was almost as a formula, with strict logic. But he missed the part of the formula that calculates how memories are constantly re-created using the machinery of our mental time machine. William's younger brother, the well-known author Henry James, so looked up to his intellectual big brother that he endeavored to write his biography after his death. The result was instead two volumes of Henry James's own memoirs, a look back that dealt with himself as much as his brother. What he discovered in the process was the true nature of memory:

"To knock at the door of the past was in a word to see it open to me quite wide—to see the world within begin to 'compose' with a grace of its own round the primary figure, see it people itself vividly and insistently," he writes, not unlike the way Linn Ullmann describes her relationship to her memories in *Unquiet*.

While William James described memory in a strictly mechanical way, his younger brother Henry managed to describe it far better in his literature. And yet he believed a lowly author's perception of memory was nothing compared to the psychologist's strictly scientific analysis. That the memory was *composed with a grace of its own* is, however, a far more apt description of memory in the way we understand it now. The insistent vividness Henry James describes is a result of the reconstruction our brain engages in when we remember something. It is a creative process that happens without conscious effort. Today's memory researchers agree much more with Henry than William James. And Henry James was not the only author to describe how the mind behaves using the meanderings of their own memory. When Marcel Proust began to write his four-thousand-page novel in 1908, he based his writing on the very nature of constructive memory. The novel grew out of Proust's spontaneous memories, his sudden recollections of times past. His life's greatest work came into being not as a traditional story normally unfolds, but as something that swelled little by little as his memories appeared. To write based on memories is to take the process of remembering and transform it into words. The border between creative writing and brain processes is vague. Writers are the most visible representatives of memory and inspire both researchers and ordinary readers to reflect upon its nature.

We will never relive our experiences as they played out in reality, but our memories can, in theory, stay with us until we die, albeit distorted, beautified, reconstructed, elaborated. One of the leading memory theories now is that the hippocampus assembles the elements that make up a memory like a director of a play. When we reach for a memory, the hippocampus finds all the elements and arranges them for us, filling in any missing details from what else we know of the world. As we remember something personal, we are simultaneously adding it to our life story and elaborating on it as an episodic memory.

During psychology's humble beginnings, it was difficult to conduct memory research. Memories are subjective, and examining them was seen as unscientific. Still, the first psychology researchers tried to make the study of mind and behavior a science, like physics. Just as in physics, that study involved measurement. Memory was reduced to something that could be quantified—a set number of words, for instance, that entered the brain and once forgotten were subtracted from the total. The memories these researchers studied were as sterile and impersonal as possible; they didn't want personal memories to get mixed in and dirty the evidence, so to speak.

The unpredictability of personal memories meant they stood in the way of science, which was rational and controlled. Only for the past twenty to thirty years have personal memories been really considered a field of interest to memory researchers, and for the past ten, the field has flourished.

According to Dorthe Berntsen, we can thank the introduction of modern brain imaging techniques, such as fMRI, for this increased research interest.

"'How can we be sure someone has a memory of something if we can't verify it?' That's what we thought about personal memories before. But now we can measure those memories and see whether what happens in the brain reflects what the test subjects have only been able to describe to us before," she says.

With fMRI we can *see* the memories acting in the brain. Even though the experiences remain hidden in the awareness of those who remember, the patterns that light up in the brain images that accompany the memories are measurable. And the patterns that appear are fairly consistent. When people allow a personal memory to play out across their inner movie screen, we see simultaneous brain activity across areas both in the front and back of the brain, and in the hippocampus, in a coordinated pattern.

"Memory research is perfect for functional MRI studies. It can be carried out without any special equipment— beyond the MRI scanner—and the research participants perform a mental activity they are highly familiar with, without any outer influence," Dorthe Berntsen explains. "All we have to do is ask the test subject to retrieve a memory, perhaps with the help of a key word."

The test subject ends up doing most of the work, as opposed to other types of psychological experiments that require chessboards and divers, questionnaires and interviews, film clips and gorillas. Remembering is such a normal part of life that we don't have to wait particularly long for our test subjects to start remembering. Researchers have actually discovered that the network that's activated when the test subject is asked to think of a personal memory is remarkably similar to the network of brain regions that is activated when one asks people not to think of anything at all! We call this the *default mode network*, as opposed to *task-dependent networks* engaged when test subjects are asked to use their brains actively to solve a math problem or count backward. In the default mode, memories pop up easily. Who can really think of *nothing* when asked not to think of anything in particular?

Though fMRI makes memories visible and measurable, it doesn't mean that brain scanning is the only way to make memory "scientific." It is high time now to insert a small warning about brain imaging: just because something lights up in an MRI doesn't mean that it's real.

When used incorrectly, fMRI has even shown signs of empathy in a dead salmon. The salmon was placed in an MRI machine by four American psychologists, who then "instructed" the salmon to look at several situations from a different perspective than its own. The results were surprisingly good. An "empathy center" lit up as an unmistakable red dot in the brain of the dead fish—but this was because of the way the results were read, not something actually happening in the salmon's brain. The researchers conducted the study to make others aware of some important principles regarding proper use of fMRI, and ways the statistics and mathematics behind fMRI can be misapplied. They received the humorous Ig Nobel Prize for having demonstrated that even a fish ready to be served for dinner can show signs of compassion, if you look for it hard enough.

Used correctly, brain imaging is a good tool for showing how memories are organized in the brain. Although we don't need MRI to tell us that memories *exist*, the mapping of brain functions to brain regions and networks is helpful in the quest to solve the riddle of memory disorders caused by Alzheimer's, epilepsy, and other diseases.

Research on personal memories is also carried out *without* MRI machines. Seeing the signs of a memory in a magnetic resonance image is not the same as understanding its content. It's like looking at a vinyl record as opposed to hearing the music recorded on it: understanding how the track makes the sound does not give us the sound itself.

The closest researchers can get is to ask people to describe their memories. The ways people experience memories vary, and your experiences are unique to you. You alone have the blueprint of your own experience. Solomon Shereshevsky's memories were colorful in a completely different way than yours and mine, since he had synesthesia. Those of us whose senses aren't entangled cannot imagine the way he experienced the world. We can only take his word that he was telling the truth when he described his recollections to researcher Alexander Luria.

How events recorded in a memory took place may be the simplest part to figure out. Descriptions of what happened, however, are often riddled with flaws and missing details—something we'll return to in the next chapter. But what is most difficult to capture is the texture of memories: what sensory experiences they contain, what feelings flow through them, how intensely they come to life. We often use questionnaires for this, in which people rank their own memory experiences on a scale.

Why do we make this our business? "Research on personal memories helps us understand, among other things, depression and post-traumatic stress disorder," says Dorthe Berntsen.

This is one of the most compelling reasons to research people's personal memories. One of the discoveries that's been made is that individuals suffering from depression have less distinct personal memories. Learning more about why can help depressed patients better hold on to their memories, so that they can find joy in experiences from the past. A study done by Susumu Tonegawa and his research group showed how happy memories can actually relieve depression. In an experiment involving mice, they first determined which neurons were activated while a mouse had a positive experience—in this case, a pleasant encounter with a mouse of the opposite sex. Then they stressed the poor mouse for ten days by immobilizing it for two hours each day, until it was fairly depressed. That's when they delivered the surprising antidote: they reactivated the mouse's positive memories by "turning on" the network of neurons connected to the original positive experience. After a few days of this "therapy," the mouse was fine again; it became more active and showed a new interest in its surroundings. By comparison, mice that were offered new pleasant moments with a partner of the opposite sex did not get any less depressed. Reliving a memory was more effective than experiencing something nice in reality. Maybe this proves that a happy memory is our own built-in antidepressant?

We think of our emotions as irrational and volatile. They don't follow logic, and they can disappear before we have time to put a name to them. They flow through us organically, coloring our lives and our memories. They're the joyous amazement that simmers when you see two seahorses swaying in the seagrass in Madeira. They aren't columns and graphs and white lab coats. How can we force emotions inside the scientific framework that makes research possible? For years, researchers in laboratories have tried to isolate the effect emotions have on memory. It's like using a chemistry set with flasks and test tubes, where the goal is to distill the purest form of emotional memories. Imagine the following student assignment in a natural science lesson:

Creating Sad Memories

Equipment: Film clips of natural disasters, children sick with Ebola, funerals, and tears. Some willing volunteers. A questionnaire about a personal memory.

Procedure: Place each volunteer in front of a computer monitor and play a film clip. Follow the changes in the person's face, from a neutral, perhaps a bit curious, facial expression to more and more sagging at the corners of the mouth, then a concerned frown and increasingly moist eyes.

Then give them the memory questionnaire.

Results: The past suddenly has a grim feel to it. Repeat until the teacher is satisfied with the number of sad memories.

But what happens if we don't have any personal memories? We don't mean like Henry Molaison, who couldn't remember anything that happened after his surgery. What if we *know* what took place, only the episodes won't let themselves be recalled, can't be reenacted, aren't coming alive? For Susie McKinnon, the memory theater is on strike, or maybe never actually opened. She lives in Olympia, Washington, and has been diagnosed with *severely deficient autobiographical memory*, the first person in the world to be identified with this disorder. This means she can't recall a single episodic memory from her own life. She knows she is married to Eric Green but isn't able to recall details from their long marriage, or to relive those experiences in her mind. She can't even recall how they met in a bar in the 1970s—only her husband can describe that memory in detail. She knows they have taken many fantastic trips, but she can only point to the souvenirs on display in their home if anyone asks about the Cayman Islands, Jamaica, or Aruba. She knows who the people are around her, though, and has successfully worked in health care and as a pension expert for the state. Her entire life, she has been a well-functioning employee, wife, and friend. There is nothing wrong with her semantic memory.

The difference between semantic and episodic memory was first described by Canadian researcher Endel Tulving. Semantic memory is, as we've said, what you know about yourself and the world—these memories are the facts about our lives which we feel we know are true. Episodic memory is what we experience when we travel back in time and *find ourselves right where something happened.* We conjure smells and sounds and feelings, and whole scenes are reenacted in our minds. It's our inner time machine that lets us feel and hear and taste and see things that are no longer here.

Susie McKinnon is missing all of this. However, living a quite ordinary life with no obvious problems, it was easy for her to believe that everyone felt the way she did. In an interview with the *Huffington Post*, she describes the first time she was given a psychological questionnaire; she was confused when they asked her about childhood memories. She believed nobody could remember their childhood and thought everybody was doing what she was doing: making up episodes from the past to spice up conversations. She didn't realize there could be something special about her memory until she read about Endel Tulving.

For years, researchers have documented people who lack episodic memory, but these individuals were usually victims of injuries or traumas that made their memories dysfunctional. But Tulving anticipated that someone like Susie would come along one day; he assumed that it would be possible for many people without episodic memory to exist in the world, undiscovered, simply because they can live full lives, have distinct personalities, and keep jobs and families, all without having episodic memory. Psychology professor Brian Levine, who studied Susie and others with her predicament, found that it's more common than previously believed. Via an online questionnaire about memory, he has—so far—received more than two thousand responses from ordinary Canadians.

"Many of them report severely deficient autobiographical

memory, so I'm beginning to think that this is not a rare occurrence at all."

"Is it normal not to remember anything from childhood?" the writer and musician Arne Schrøder Kvalvik asked us when he realized we were writing a book about memory. "And I mean nothing at all!"

Arne was nominated for the Norwegian Brage literary award for his first book, a nonfiction work called *Min fetter Ola og meg* (Me and my cousin Ola), and is part of the award-winning band 120 Days. He is a respected musician and writer, and a devoted father. But something is different about him. Like Susie, he remembers nothing from his years growing up.

"Gradually, I guess I realized that I was different, because people were mentioning what we had done as kids. But I remembered nothing," he tells us.

He knows very well who his relatives are, where he went to school, what he did in his spare time. He just doesn't have any memories of those things. He can tell you where in the school his classroom was, but that's about it. He has nothing beyond the location. He is unable to recall how it smelled, what he said to his teacher—neither happy nor sad memories appear. He knows he played with his band in front of the whole school, but he can't recall the experience of standing there, young and nervous, with classmates and teachers watching him onstage.

"I worried that I was repressing memories, that I had experienced something traumatic. But that's not it either! My memory is simply weaker and weaker the further back

I go. Things that happened ten years ago I can remember, but only a few episodes—even though I was traveling around the world as a musician. I have performed in concerts throughout Japan and the USA."

His life doesn't lack unique moments that should have etched themselves into his memory. But it doesn't bother Arne that he remembers so little. And if we are measuring people by their success, he has it all: he is a happy dad of two with a brilliant career. But behind him, the memories are waning, disappearing, evaporating into thin air. When his parents told him that he had been in Portugal in his teens, no pictures emerged from his memory of the family they shared a house with, of the sunlight, the sea, the bacalao, or old, dilapidated stone houses. He remembered only one thing: a white pair of pants.

"I don't remember my first kiss, for example—and not because I was drunk. But there is one thing I remember from when I went to school: the terror attack on September 11, 2001, when I was seventeen," he says. "But I only remember where I was when I watched it on television, nothing else."

He doesn't remember how he met his love, or what they talked about during those first trembling moments of romance, but he knows he loves her. And, after all, that's enough. "It's not a problem for me that I have so few memories. It doesn't bother me."

We don't know why it's like that for Arne. We don't know if he falls into the same category as Susie McKinnon, who has her diagnosis—if we can call it a diagnosis. Perhaps it is only a reflection of the great variety in people's memories. Some experience memories very visually, others don't. The opposite of severely deficient autobiographical memory is what Levine and his colleagues call *highly superior autobiographical memory*. People with this memory capacity often remember the exact day something happened, even many years later, and their strong feelings connected to past experiences don't wane easily.

Ida Jackson, the blogger we met earlier, is the complete opposite of Arne. To her, the ability to remember *is* important. She wants to remember everything, and most of her memories appear as evocative images—sometimes too colorful.

"Sometimes my memories are so strong and unpleasant that I feel physically sick, and to work through these feelings, I talk to a psychologist and write in my diary and on my blog. Writing about a memory sort of removes its power; it simply turns down the volume of the memory. In that way, I change the memory from something I see and hear before my inner eye to a story," she says.

One of the memories she has written about is her experience of being bullied in school. Her blog post was picked up by a magazine in Oslo, which posted Ida's story on its website long after her blog post had first appeared online. Six years after she'd first written it, Ida watched her post go viral on social media.

"On the internet, it's not newsworthiness that determines whether a post is shared—it is the emotional strength of the writing. This blog post was very painful for me to write. I shared all the shame I felt with my read-ers. The problem is that my memory was highly biased by a story about being a sympathetic victim, but that was not really the case. I smelled bad and picked my nose and ate it in front of my classmates. But that did not fit into my story, so I didn't include it. I didn't deserve the horrific bullying I received, but in truth, I was revolting! It was up to the adults to stop the bullying, of course. Beware, that when we write stories based on our memories, we often grab the most stereotypical tales available. When I saw how my blog post swept across social media, I had to correct myself and write a truer story."

"We are strongly affected by the story lines in Hollywood films, we know that. We look back at our childhoods and try to find foreshadowing of who we will become: the sign, the key, the trigger that makes things what they are," clinical psychologist Peder Kjøs says. He participated as a group therapist for youth on a Norwegian TV documentary in the spring of 2016 and writes a psychology column in an Oslo newspaper.

In his therapy practice, Kjøs helps people restructure their lives by writing their life scripts. Our memories become stories about ourselves. Some stories are easier to hang on to, because they fit our self-image better than others. In many ways, the psychologist is a coauthor of life scripts, or at any rate a cautious editor. And we are all authors of our own life stories.

"We have a tendency to look for a certain dramatic struc-ture in our lives. Since we can't scroll forward in our lives, we look backward to write the story about ourselves. And when we rewind, we direct, cut, and photoshop. We can change the script as we go along, find reasons for things being as they are. Sometimes, clients want there to be a turning point in their childhood, so they don't have to blame their parents even if they were neglected. On the other hand, if we have had it reasonably good but life is not going as expected, we may well blame our parents. Of course they may have done something wrong or fallen short; no parents are perfect. This doesn't mean that people haven't had bad experiences during childhood, but we often lend great significance to modest events."

His clients' problems arise when they edit their stories too harshly. It's one thing for our memories to be reconstructive and elastic, and another if we bend them around absolute untruths.

"A narrative that doesn't agree with facts doesn't work. It is important, however, in a treatment situation for clients to arrive at narratives themselves. I don't know what's true, and I can't present them with a solution."

Working with people whose life stories are made up of predominantly bad memories demands something special of the therapist. It is important to give the clients a feeling of power and responsibility without saying all those bad things that happened were their fault. Without power and responsibility, we can't change anything—we become a minor character in our own narrative.

Most difficult is changing a life that has only gone from one dark moment to another. How can we imagine anything good can happen if all we know from the past is pain? How can we see the light at the end of the tunnel? In *Soria Moria*, a painting by the popular nineteenth-century Norwegian artist Theodor Kittelsen, a hero from folk tales gazes toward the glow of the sunrise, which resembles a golden castle on the horizon: it's a mirage, a goal for Askeladden's wanderings. Askeladden's courage and daring are probably what allow him to see a castle at all, just beyond the dark rolling hills. To a depressed person, even the hills would be out of focus.

THIS BRINGS US back to the lush forest landscape over which our sister has finally jumped out of the plane, parachute fastened to her back. Her heart beats quickly and the adrenaline pumps so vehemently through her body that it's hard for her to perceive what's happening. She releases her parachute after ten seconds of free fall. Contrary to what we might believe, the most dangerous part of the jump is still ahead of her—the landing. The much-anticipated feeling of flying is all-consuming. She is overwhelmed by it and by the routines she must keep in mind throughout the jump. She completely forgets to pay attention to the ground rushing toward her. At first, it seems like an abstract map, a quilt of forests and fields. The trees speed closer and closer, and every second counts. The field she was supposed to land on glides past below, and she is on her way toward the forest and what might be her return to the ground. Finally, she reacts and steers the parachute. But it is just a little too late!

The top of a tall spruce tree is where Tonje ends up.

Though unharmed, she dangles there for two and a half hours before the search team finds her. And yet, this experience does not deter her from trying again soon afterward. It is not until she is sent up with the plane to make her second jump of the day—her second jump alone, ever, actually—that she has an experience we expect on our deathbeds: her life passes before her eyes. It turns out that the images are disappointing! Instead of high point after high point, pearl after pearl of unforgettable moments, she sees only meaningless, mundane images from childhood, mostly standing on the lawn at home as a seven-year-old, or walking, backpack on her back, along the asphalt driveway up to our house.

"They were very boring experiences, and I don't know why they were the images that popped into mind," Tonje says now.

The second jump cured her of the anxiety she felt from the first jump. Since then, she has jumped thousands of times and never experienced anything like it again.

We assume that as we die, the most important moments of our lives will appear before us. That's why we collect them, isn't it? Suddenly, as we're about to write our life story's finishing line, our most important memories are supposed to present themselves to our inner eye. We will see clearly what life was all about. The actual meaning of what we've experienced will be revealed to us. The end of the story will explain the beginning. Or will it?

Caterina Cattaneo, who led our diving experiment in the Oslo Fjord, was unconscious under water—medically defined as drowned—for thirty minutes once. After a particularly difficult dive, Caterina woke up in a hospital in Oslo. She had drowned. She was only twenty-one years old, and she had taken risks in the water. The last thing she thought as she was embraced by the darkness: *That guy at the party last night, why didn't I sleep with him?* None of her important memories surfaced; no crucial revelations. Her life script was neither rewritten nor revised. There was just a trivial thought, and then black.

The last thing Adrian Pracon thought at the cliff's edge, imagining he would die, was more concrete. He envisioned the coffin containing his body being lowered into the ground. His parents crying, shattered by grief. The image came to him from nowhere, without his willing it to come, and he was surprised at how clearly he saw it, as a murderer aimed a gun at him.

Adrian Pracon was twenty-one years old and a youth politician. For the first time, he was participating in the Norwegian Workers' Youth League summer camp on Utøya, an island in a lake about half an hour's drive from Oslo. Some six hundred committed young people from around the country had gathered for a few intense July days of political work and evenings with singing and political discussions. Even the country's prime minister and foreign minister, as well as previous prime minister and former World Health Organization director-general Gro Harlem Brundtland, visited the camp that summer in 2011. Politics, power, and youthful dedication came together on that small island, coloring the nights with hope and plans for the future.

Anders Behring Breivik had been preparing for a long time. The bomb at the government building, which killed eight, and the attack on the youth camp, where he killed sixty-nine, were thoroughly planned in solitude. As Adrian envisioned his own funeral, the killer stood on the shoreline at the southern tip of the island, aiming his gun at him. There he stood for a couple of seconds. Then he lowered the rifle and walked on. Adrian remained hiding in plain sight, lying down on the point, which jutted out like a gray finger over the water. The landscape was completely open and unprotected, rocks dotted with low, naked bushes. Adrian lay there with a jacket over him, pretending he was dead. Perhaps doing so saved his life when the terrorist returned to the point to make sure everyone was dead. Failing to realize Adrian was still alive, he may not have bothered to aim properly. Adrian was hit by the last shot fired on Utøya. The bullet entered his shoulder and exploded into more than seventy pieces of shrapnel, lodging in the muscle tissue as little reminders of the day his life changed forever.

"The first years after it happened were totally out of control. The memories returned completely involuntarily, often when I was stressed. The more I needed to clear my head, the worse it got. Three years of my life are completely lost," he says today.

He survived the shot in the shoulder, but life after the massacre was not the same. Adrian had been on the threshold of something new and exciting; it had felt as if he were soaring into the future. He had landed the job as regional leader in the Telemark Workers' Youth League; he had a boyfriend, a dog, and a place to live. But after the nightmare on Utøya, his life became a constant repetition of what he had gone through. Over and over again, he suffered through some of the most horrifying moments of his life. He analyzed them from all perspectives, thought of what he might have done differently, how little it would have taken for the bullet to enter his head, his heart, or his spinal cord, rather than his shoulder. Sometimes he would imagine picking up a rock and killing Breivik before he could do more harm. Sometimes he would focus on his guilt, because it had been up to him, as regional leader, to recruit participants for the summer camp. One of the youths he recruited never returned. He was only fifteen years old.

In the period after July 22, Adrian began to drink heavily. Uninterrupted sleep was a luxury he could no longer enjoy. The man who used to be boyishly messy became extremely fussy about cleanliness; his behavior bordered on mania.

"My boyfriend at the time told me I would take a sleeping pill and then clean the whole house. I remembered nothing and woke up to a clean house."

During Breivik's prosecution Adrian began to write a book about his experience. He drove back and forth from his hometown of Skien to Oslo to follow the legal hearings and work on the book. It was a very hectic and emotional time. One afternoon, after a couple of beers with friends, everything went black. Adrian came out of it as the police were handcuffing him. He had to go to court himself, and he was sentenced to community service for his violence against two men that night.

"It scared me. I can't drink like that anymore. If someone suggests a beer, I have to take a walk first, find out how I feel. I have to know I'm having a good day."

The terrorist is serving a life sentence, and so are Adrian's memories. He takes them out for fresh air and walks them in the exercise yard before he forces them back into the dark; they are strictly supervised through most of Adrian's waking hours. Allowing them out for an interview will result in a bad day afterward. Now that he knows this will happen, he can plan accordingly.

"For a period of time, I drank a lot, probably because I didn't want to remember anything at all. I didn't want to remember all those evenings when I was not feeling well. I just wanted to get away."

July 22, 2011, was a national trauma. The new realities of the world reached the safe country of Norway. The Norwegian relationship with safety and danger was changed forever. The terror also gave Norway a new memory milestone. All Norwegians have a memory tied to July 22, 2011. American researchers have called this phenomenon, whereby strong emotions tied to a shock glue an experience to our memory, a "flashbulb memory" because the experiences seem frozen in time, like a photo in which the flash has gone off and thrown a glaring light over everything in an otherwise dark room.

In many psychology textbooks, the explosion of the space shuttle *Challenger* is used as an example of an event that could have caused a flashbulb memory, as are the terror attacks of September 11, 2001. Before Breivik, Norway didn't have any obvious, tragic examples. But there were positive ones. Where were you when Oddvar Brå broke his ski pole? The time a Norwegian skier came close to losing a gold medal because his ski pole broke is a memory common to many Norwegians, one they can laugh over and share stories about (although most of the memories involve sitting on the sofa in front of the TV or standing along the ski trail during the World Ski Championships in 1982).

"Where were you on July 22?" is a question that ties our personal life stories to that of Norway. For those directly affected, it's another story; memories of what happened will follow them the rest of their lives. At the Norwegian Centre for Violence and Traumatic Stress Studies (NKVTS), the survivors and their families have been studied. When something this serious happens, it's necessary to bring in investigators and terror experts so society can learn from the events. But for the researchers at the NKVTS, the important thing is to learn about trauma, to better help survivors the next time something happens. People go through trauma all the time. They are victims of rape, assault, car accidents, and wars around the world. Many carry traumas that get far less attention than those resulting from July 22, although they are no less serious for those affected. How can they be helped to escape their intrusive memories?

Ines Blix is one of the researchers at NKVTS studying this national tragedy. She has followed those who worked in the government building where the bomb went off, using both interviews and questionnaires to see how their lives have been affected by the terror act.

"There are two traditions within trauma research about how we remember traumas and what trauma memories do to us. This has been called the memory wars. Some believe that we remember traumatic events quite differently from other events, and that traumas can lead to fragmented memories, extreme repression, and dissociative personality disorders. Other researchers, like myself, believe that traumatic events, like other emotional events, are often remembered very well. Memory behaves to a great extent as it always behaves, but it turns up the volume to an extreme degree; an ordinary memory system on volume ten."

Blix's research shows that among the most common problems people report after traumatic events are intrusive, detailed memories that appear over and over again for a long time after the event.

We have known about intrusive, involuntary trauma memories for a long time. After World War I, these were a symptom of what was called "shell shock," and in Virginia Woolf's classic *Mrs Dalloway*, they were what made a young soldier jump through a window to his death. At that time, post-traumatic stress disorder (PTSD) was an unknown illness, and people wondered why soldiers became unfit for battle without being visibly injured. They retreated from society, didn't sleep, didn't eat, were unable to take care of themselves, and often lapsed into apathetic staring or behaved in a panicky and irrational manner. The gruesomeness they were exposed to during the war was extreme. For the first time, millions were mowed down by machine guns in muddy trenches, in a brutal war that divided Europe. Since then, knowledge of trauma psychology has increased. It wasn't laziness or head injuries that made the soldiers ill. What was it, then?

A strong tradition within the research suggests, as we've pointed out, that trauma memories are different from ordinary memories. When people develop split personalities or other dissociative disorders, these are survival mechanisms that help them master their lives through the crises.

But how can that be, when we know that shocking images and strong emotions associated with upsetting events can slash memory open and burrow in far more strongly than anything we experience in everyday life? Trauma ties itself to memory in every way possible. Traumatic events are powerfully emotional, unique from everything else we have experienced, and upset much of what we assume about ourselves and the world. It's not only that trauma is hard to forget, it's that it pops up unannounced, like a jack-in-the-box. Victims of trauma are often unable to lock that box, and their upsetting memories appear over and over again in all their cruelty.

On a questionnaire given to 207 government employees who were at work on July 22, half responded that they were still troubled a year later by recurring memories from the terror attack. For a quarter of those examined, the effects were so serious that they could likely be diagnosed with PTSD. Even employees who were not at work that day were affected. They knew that they could have been at work and suffered from recurring thoughts about colleagues who had been hurt.

PTSD develops gradually over time after a traumatic event. When the memories don't subside, a person's attempts to avoid them only strengthen them and make them uncontrollable. Someone suffering from trauma avoids everything that reminds them of it so as not to relive it again and again. This disrupts everyday life and makes it harder for them to return to work and studies. It is also extremely difficult to keep from thinking about the trauma. It's like saying "Don't think of an elephant!" What happens? The elephant stomps around, knocks things over, and takes up an enormous amount of room the more you pretend not to see it.

Spontaneous and pleasant memories pop up out of nowhere too. As Dorthe Berntsen found in her research, we can be reminded of them by whatever we're experiencing at the time. Music we hear on the radio transports us back to when we were fifteen. We don't give this much thought. It becomes obvious to us only when the volume on spontaneous memories goes through the roof, as it does with traumatic memories. Obviously, these memories take up space, as does any memory. Sometimes, the memories consolidate and turn into PTSD. Other times, even the trauma memories subside and become stories that can be told without strong feelings—no longer the elephant in the room.

"The big question is why some people develop PTSD and others don't," Ines Blix says and points to several explanations that together can perhaps show part of the picture. It may be due to differences in how people handle an event in their working memory: to what degree they are able to filter undesired information out, how flexible they are in steering memory. Small variations in such basic brain functions, not noticeable normally, can make a difference in extreme situations, which trauma and life thereafter are. There can also be differences in how people organize their memories, which can result in traumatic memories taking up more space in some than in others.

Blix calls it centering. "In our study, we found that those who placed the events of July 22 in a more central place in their autobiography, as a major turning point in their lives, were more likely to develop PTSD. Centering was a predictor for who would have PTSD symptoms three years after July 22. We believe centering makes the traumatic memories more easily accessible. They become a point of reference."

It's as if you are actually riding the elephant. Not thinking of the elephant is difficult when it is constantly following you around. Then, one day, you become the elephant. You are identifying with the trauma, and it has become a part of you. It has become a central point of your life's story.

The hippocampus plays an important role here too. Several studies have found that people with PTSD have smaller hippocampi than average. The question everyone has asked in response is obvious: "Are psychological traumas harmful to the brain?" The stress reaction, when we are extremely scared, can trigger the body to produce sky-high levels of the hormone cortisol, which in large amounts can be harmful to the brain, especially the hippocampus, which is about as vulnerable as its namesake seahorse. But a unique study of twins, conducted by Mark Gilbertson and colleagues, gives a possible alternative explanation. Among the twins who were studied, one in each pair had been exposed to a psychological trauma. That's how they could compare the hippocampus—normally very similar in identical twins—in one person that hadn't experienced trauma and one that had.

What was surprising was that Gilbertson and colleagues found the hippocampi were fairly similar in the twins, both with and without trauma. "This may mean that the size of the hippocampus before the trauma can be a risk factor," Blix says.

It is still a mystery why *smaller* hippocampi can produce such powerful memories that they can almost put a human out of action. Shouldn't it be the other way around, that a *larger* hippocampus can more easily replay the bad memories?

We can't do anything about the size of our hippocampus. We can't strive to shape our memory in a way that makes us better equipped to survive catastrophe. But when the trauma is a fact, is there anything we can do? Blix believes that, more than anything, knowledge of normal brain reactions can help people handle traumatic memories better and faster.

"It is beneficial for people to know that it is common to react with intrusive memories and that for most people, the memories will weaken as time passes." The elephant will stomp less frequently around the room, which is our awareness. Eventually, we can take control and lead it back into its enclosure. It is the fear of it always being on the loose that hurts us.

Trauma treatment is first and foremost about turning down the volume on memories and breaking the vicious cycle of avoidance. PTSD puts us on high alert for spotting possible new dangers; we are easily startled and sleep poorly. Some studies even point to a worse general memory in the wake of PTSD. It may not be an odd thing to suggest, if the traumatic memories are taking up space. The fear of the memories can almost amount to a phobia. A phobia sets us up with a pattern of reactions that steer our actions. Phobias can be extremely tough to shake simply because avoiding something feared offers such relief. The relief acts as a reward. If we can choose between feeling relieved—by, for example, hanging out in our apartment, feeling safe— and going out and being reminded of something bad that happened and feeling extreme fear, it's easier to choose relief. The result is that we stay home more and more. So do the memories. We learn that the memories are to be feared, and we should avoid them. The more we avoid them, the stronger they get. The same happens if we're afraid of wasps, injections, sharks, dogs. If we avoid them, we will undoubtedly continue to be afraid. The alternative is to face our fears. How is this best done when it hurts so much?

"Trauma-focused cognitive behavioral therapy and eye movement desensitization and reprogramming (EMDR) are the preferred treatments for PTSD," says Blix. The idea is not to throw ourselves at the memories we fear like a kamikaze pilot, but to approach them cautiously and gain more and more control. We get used to the memories and deplete their powers.

Therapists often use relaxation techniques such as EMDR, in which they move their hands back and forth in front of their patient's face. It might sound weird, but it gives the patient something external to focus on while they talk through their memories, dividing their attention between their traumatic feelings and the therapist's strange hand movements.

In an ideal world, there would be a vaccine for PTSD, so that if we encountered something horrific, we could go to the doctor, get vaccinated, and immediately feel safe— not unlike getting a tetanus shot. Emily Holmes's research team at Oxford has tried this exact thing. They believe that playing *Tetris* in the hours following a traumatic event can considerably reduce the occurrence of intrusive memories. The way they came to this conclusion was by showing the volunteers in an experiment a very traumatic film. Some of the participants got to play *Tetris* afterward, while others were left to their own devices. Then all the researchers had to do was wait for the traumatic memories to appear—voluntarily, that is. In these experiments, *Tetris* had an obvious protective effect. The idea is that the game competes for space with the strong visual memories associated with the trauma. The visual impressions immediately following a traumatic event remain vivid as they begin consolidating into memories. Playing *Tetris* prevents the images from the trauma from having proper access to our memory. In comparison, they found that those who played a word game—such as a quiz—had more flashbacks than those doing nothing. The linguistic distraction probably kept the participants from storing their own interpretations and assessments of what they had seen, but it still left them with a set of immediate, intense visual impressions.

Will this therapy work in the real world? When your world has been turned upside down, and you're thrown into a completely genuine panic state—not just a reaction to a film in a lab—will it seem reasonable to pick up the smartphone and play *Tetris*?

What may feel more natural to most of us is to try to understand, find a context, some order. The experience of a traumatic event is not like a film. Our attention decides what makes it into our memory and what is left out. But our attention is also affected by terrible fear. We cannot take everything in. In the experience of trauma, our personal worldview—what we normally use to interpret and understand new experiences—is put to an enormous test. Something that breaks with all expectations of a peaceful life, such as a bomb in a government building, takes an extra long time to understand. We don't have time to understand it at the moment the bomb explodes—the understanding will come later, perhaps never. When researchers at NKVTS and the University of Oslo examined the stories of the youth that survived the Utøya massacre, they found that those who developed PTSD symptoms remembered more external details about the event and had fewer internal thoughts and interpretations. This suggests that those who continually assess and interpret their situation are better prepared to handle their memories, while those who are attending to the details of their surroundings are more troubled by their memories afterward.

Adrian Pracon worked through all the details from the horrific day on Utøya when he wrote his book, *Hjertet mot steinen* (Heart against stone). Writing helped him to put the gruesome details behind him. Now he doesn't feel the need to remember as much. His detailed memories are contained within the black box—his book—and don't fight to get out as aggressively as before. But making the memories fade wasn't as simple as writing a book; that was just one step on the path toward his new life. Like everyone who has been through trauma, he longs for the normalcy of everyday life. The PTSD always has him on high alert. In new places, like a café, he will still look for possible escape routes and hiding places, and the screams of children playing can make him nervous. Seeing pictures of the murderer inevitably triggers a reaction. Of course, it is almost impossible not to see pictures of him. Especially in the time following the massacre, Breivik was all over the newspapers and blogs and incessantly talked about on TV and radio.

"I saw him, once. It was at the convenience store. He stood in a corner and turned toward me. I had to calm myself down; I knew of course that he couldn't be there."

A picture or a comment, a poor night's sleep, or something else had triggered the memory of Breivik so strongly that he was visible to Adrian. Several survivors state that they are thrown into the situation and *are on Utøya* when they hear the loud voices of children. They can see the grass below and the trees around them and feel the panic reaction take over, even if they are safe, in the middle of the city.

Returning to where it happened may be the most frightening thing Adrian could do; the memories are extra strong on Utøya. That's part of what we showed with the diving experiment: how a memory is connected to, and appears in, a particular place. What, then, do we want to accomplish by bringing Adrian to the place where he was shot? Won't seeing it again overwhelm him with bad memories?

"It's beautiful, but for me, it's like there is a dark cloud over the island," he says as the boat, MS *Torbjørn*, steers toward the pier. The chugging sound of the old boat's engine fills the April afternoon with summer memories. The water rushing along the side of the boat may tempt us to go swimming, but it emits a raw chill. On July 22, close to five years ago, it was raining and cold here. The Tyrifjorden—one of the deepest lakes in Norway—wasn't suitable for swimming, even in the middle of the summer. When Adrian stood on the shore that day and felt the water fill his shoes, he suddenly woke up and realized that something very dangerous was happening. Even a few minutes earlier, when he watched a young girl get shot, all he could think was *It must be an exercise, role-playing.* It wasn't real.

"It still controls much of my experience of what happened, that I thought it wasn't real."

We walk around the island, which is now dotted with spring flowers, on a kind of guided tour of terrorism. Had we been simple tourists, we wouldn't have known it was anything other than a beautiful idyll. Blue and white anemones, typical Norwegian spring flowers, poke up everywhere between rocks and under slender trees. And even though we feel like it, we are no more voyeurs than those who go to Auschwitz to comprehend more of the Holocaust. It is sad and strange to be here.

New buildings are being erected here to replace the café where many of the bodies of those killed were found; they will become a learning center for democracy and freedom of speech for young people. We are being shown around one of the buildings by a manager on Utøya, Jørgen Watne Frydnes. Inside the newly erected building, which includes a café, an auditorium, and large windows that frame the nature outside, there is a library of political literature. The shelves are fifteen feet high. We remain standing for a long time in front of the bookshelf. Here, the politicians of the past will speak to the politicians of the future from the pages of books.

Right by the new construction, somewhat hidden in the forest, there is a memorial to the sixty-nine who were killed here, a delicate metal circle suspended from the trees. The names of the victims and their ages are stamped into the metal, children who should be here and would have been five years older today.

Adrian stops at the circle and reads the names. We put flowers on the ground under it. Then we walk down toward the place where Adrian was shot. A swan floats in a small bay, illuminated by the sun that emerged after the spring snow that was falling when we first arrived. The fragile flowers have captured white snowflakes that already have begun to melt.

"I don't think there was any real change in me until I accepted that life as I knew it would never return. That's when I quit fighting and could begin building my new life, the one after the event."

The memories can still overpower him, but he has gained more and more control over them. Trips to Utøya (this is not his first) serve as yet another way to gain the upper hand.

"I know from experience that periods of stress release the memories. This can ruin an exam, a job interview, an assignment deadline. This winter, I was stupid enough to wait until the last day to finish a take-home exam. Then the trauma memories took control in the middle of all the stress, and I couldn't deliver on time."

Adrian has already planned not to do anything after our trip to Utøya. Taking the memories out for a whole day in the fresh air is demanding enough. "Much about me changed after what happened. I was messy before, now I am orderly. I wanted to work in the civil service, now I want to research terrorism. I didn't used to read the foreign news in the paper, now that's all I read. Before, I couldn't envision myself living in Oslo."

Adrian moved to Oslo in 2012 and eventually started peace and conflict studies. He wants to conduct academic research on terror. He is hoping that what he has lived through will help inform his research.

We're gazing beyond the point, where he lay during most of the terror. It is covered in gray rocks and some dried-up, half-dead trees that even in the summer hardly have any leaves.

"I can see the ghosts of everybody who came and went here, and I can see him. It is like a film, where you see transparent people come and go."

"Have you ever wished that you could forget everything that happened before July 23, 2011—erase it from memory forever?"

"I have daydreamed about it. When I was very sick, I thought about it a lot. But I have so many good memories too. I don't want to lose them."

IN THE CUCKOO'S NEST

Or: When false memories sneak in

"I can't believe *that!*" said Alice.
"Can't you?" the Queen said in a pitying tone. "Try again:
draw a long breath, and shut your eyes."
Alice laughed. "There's no use trying," she said. "One
can't believe impossible things."
"I daresay you haven't had much practice," said the
Queen. "When I was your age, I always did it for half-an-hour
a day. Why, sometimes I've believed as many as six
impossible things before breakfast."
LEWIS CARROLL, *Through the Looking-Glass*

HOW MUCH IS it possible to believe, without a belief
being true?

This "memory" was donated anonymously to the
False Memory Archive:

"I remember so clearly taking a medal that belonged to
my father and burying it in the garden. I then looked for it

for ages, digging up little pieces of earth but never found it. When I think of it now it must be a false memory. Why would my father have medals? Why would I bury them? But the memory feels like truth—shiny colours and crisp edges."

This may not be the most serious of misrememberings. Something like this may or may not have happened—most likely not, yet it feels as real as any memory. But as innocent as this false memory may be, it illustrates how vulnerable our personal memories are and makes us question whether we can trust our own past. Artist A.R. Hopwood became interested in Elizabeth Loftus's research at the University of California, Irvine. Loftus is the world's leading expert on false memories. Hopwood decided to make an art project, the False Memory Archive, surrounding people's common distortions of the truth. People who falsely believe they have experienced an airplane's emergency landing or a car crash appear side by side with someone who is absolutely sure he remembers Live Aid in 1985, even though he was born after it happened. In the course of Hopwood's tour, his audience contributed to the combination of art and documentation by adding their own false memories, making the collection grow.

One would think that since people believe their memories are true, they would not recognize that they remember things that never happened. Nevertheless, the project has accumulated a significant number of false memories. Many of the memories date back to early childhood. Memories of having flown around the nursery are easy to explain as the result of a certain lack of understanding of reality in small children. The reality of such a memory is, of course, soon discarded when we are old enough to realize how things really must have happened. But false memories can also occur in people with a fully developed memory and a sound worldview. Psychology professor Svein Magnussen, who has devoted much of his career to false memories, was himself the victim of a false memory. For a long time, he thought he had committed a crime in his youth.

"We had traveled from Oslo to Copenhagen in the little car we had purchased for high school graduation. The car broke down. I remember clearly that we pushed it off the pier, and it ended up in the water. I even remember a wooden pier—although, when I think about it, I am sure that there are no such piers in Copenhagen," Magnussen, now professor emeritus at the University of Oslo, tells us.

For thirty years, he believed the story of the car was a somewhat off-color episode from his life; it is, after all, illegal to get rid of cars that way. Then he met his high school buddy again at a party. His friend was the one who had bought the car in the first place, and, it turns out, eventually sold it to a junkyard in Copenhagen. It had never been pushed into the sea!

"Somewhere along the way I had made up a clear memory of us pushing it off the pier. We may have discussed it; it may have come up as a possibility. And then I created an image of the situation, which wound up in my mind as a true memory," Magnussen says.

His story demonstrates an unpleasant fact: not everything we think we experienced is real. Not even remotely.

We create false memories in many ways. We may "steal" other people's memories. For example, we know that war veterans in group therapy gradually adopt each other's stories. For some, becoming engrossed in another person's exciting story over dinner can lead to that story finding a place in their own memory. Childhood memories can be even murkier—it is often unclear if a personal experience is something that really happened or just a good story that a peer group has told over and over—or perhaps something you have seen in a photo. False memories can be formed when we see something on TV, take part in group therapy, or talk to our siblings about something that happened in childhood. So can we trust our memories?

"Why some people are inclined to create false memories, and others not—what characterizes a person who creates false memories—we don't know. You would think that people who remember their lives clearly, in vivid detail, would be incapable of making up false memories, but that is not the case. Even they may carry around false memories," says Magnussen.

The unbelievable stories in the False Memory Archive ruin most of memory's credibility. Our memory is reconstructive and elastic. It is not a PDF we open and reread on a computer, or a camera filled with high-definition images. Memory is more like live theater, where there are constantly new productions of the same pieces. In some performances, the heroine wears a red dress, in another a blue one. Now and then, actors are replaced and the plot revised—even quite substantially. Sometimes the play is about something we have actually experienced, other times about something we have just imagined. In the memory theater, there are many strange reenactments.

Each and every one of our memories is a mix of fact and fiction. In most memories, the central story is based on true events, but it's still reconstructed every time we recall it. In these reconstructions, we fill in the gaps with probable facts. We subconsciously pick up details from a sort of memory prop room. This is the brain's way of being space efficient—we don't need to store everything we experience as exact film rolls. We can store items as persons, things, sensory experiences, actions—each stored on its own but tied together in a memory network maintained by the hippocampus. This frees up space in our brains and liberates our thoughts. We are not slaves to our memories, even though we actively use them all the time. But this elasticity comes at a cost: things can easily become confused. Like when a witness believed he had seen two people rent a truck the same day in 1995 that Timothy McVeigh bombed Oklahoma City, killing 168 people. This kicked off a manhunt for another accomplice—on top of McVeigh's known accomplice at the time, Terry Nichols—who did not exist. The witness, who worked at the body shop where the truck was rented, had definitely seen two men. However, he saw them the day after the terrorist had been there, and one of the two men unfortunately looked a little like McVeigh. The witness had confused the two events. He exchanged McVeigh's face for that of one of the innocent customers he saw the day after. This witness has a memory no worse than most people's. For any clerk, what customers come in at what time is not worth remembering—normally.

Mix-ups like this aren't noticeable in our everyday memories. Yet if we were to examine each of our memories down to the smallest detail and compare them to, for example, film footage, most would fall short. Visualize your office or classroom or convenience store. You probably won't remember every single detail—what books are on the bookshelf, how the cord of the cell phone charger coils across the desk, where the coffee mug sits, or how the light from the window reflects on the wall. Still, the memory can feel trustworthy enough. You have enough memories of the coffee mug and the cell phone charger to be able to pick them up in the prop room and put them in place. If you have spoken in front of a group of people, you won't remember every single face in the room. Still, if you try to conjure them from memory, the room will be just as full as the first time you were there. But though the atmosphere may be the same, the individual people are picked from a pool of extras in your memory.

In fact, according to a study conducted by Loftus and her colleagues, people with very good autobiographical memories remember more incorrect details from a picture presentation than people with average memories. It seems as if those with good memories use their memory theater to the fullest. They have a large memory repertory, but with that comes a tendency to reconstruct a little too much.

But there are more factors at play when it comes to constructing false memories. The more time passes, the more probable it is that fantasies—like the car that was pushed off the pier thirty years ago—can sneak into our memories. The time aspect is significant; we seldom recall incorrectly what happened yesterday, but what happened last year is more unclear. Even though the most entertaining examples involve dramatic events, mundane happenings have an easier time becoming false memories compared to remarkable fantasies.

Solomon Shereshevsky, the man who couldn't forget, claimed to remember what it was like being a baby. He described in detail how the light fell across the cradle, his mother—or nanny—far above him. But since he had such a vivid imagination, this is probably a false memory. It is difficult to believe that Solomon would be exempt from the rule that everything we experience early in childhood disappears into the abyss called childhood amnesia.

Solomon's imagination often played tricks on him. He had a terrible experience as a child when he and his family were moving to a new home. When they departed in the moving truck, he imagined that he was left behind, a vision so vivid that it felt real. What researchers have found, using modern MRI scanners, is that when we imagine something, the activity in our brains is similar to what it would be when we experience the same thing in real life. Imaginings, memories, and false memories actually behave quite similarly in our brains. It is only how our brain sorts things under the labels of "true" or "not true" that determines what's what. A real memory *is* a form of imagining—an imagined reconstruction. False memories make use of memory's natural laws, however irrational they seem. A false memory moves, in some way, from imagination to memory and is suddenly seen as something real. It steals a label that says "true," and just like a common cuckoo—a bird that lays eggs in other birds' nests—it hatches in the reed warbler nest, pushes the baby warbler out, and begins to grow into a big, fat cuckoo bird.

Since our imperfect memories can make us believe in something that never happened, can memory be manipulated from outside? Is it possible to deliberately create false memories in others?

Actually, researchers have managed to plant false memories in the brains of mice. Do you remember what we said earlier, about place cells in the hippocampus that help us remember certain locations? Researchers placed an electrode in the hippocampus of a mouse, where the place cells are, so they could register the nerve signal when the mouse arrived at a certain spot in its cage. Then they waited for the mouse to fall asleep. When both mice and humans sleep, place cells are activated—it's as if sleepers return to the places they've been to that day, to keep them in memory for later. And so this particular place cell in this particular mouse reactivated on its own while the mouse slept, because that's what place cells do. This is where the researchers could do some manipulating. They also planted an electrode in the reward center of the mouse's brain. When the researchers applied a tiny touch of electricity in this area, the mouse was bathed in a feeling of wellness, similar to what it would have felt if it ate sugar, had sex, or did something else that felt good. The reward center sends out neurotransmitters that, in addition to making a mouse or person feel good, also help strengthen new connections between neurons, creating a memory trace. The researchers stimulated the mouse's reward center at exactly the same time as this particular place cell was active. This strengthened the connection between the exact location this place cell represented, which the mouse had stored in memory from before, and the feeling of wellness it experienced. Links like this are usually made while the mouse is awake by giving it sugar or some other treat just when it is at the chosen place. In this experiment, however, the link was created artificially, without the mouse ever feeling good at the specified place in reality. But after having been supplied a memory of wellness at that place in its sleep, the mouse returned more often to that place in the cage. It had been given a false memory.

In a less pleasant experiment, mice were implanted with false memories of receiving electrical shocks at certain locations in their cage using optogenetics. Optogenetics uses light to control cells by implanting a gene that codes an on-and-off switch steered by a laser. Such light-driven switches exist naturally in some organisms (e.g., one-celled algae), but with the help of gene technology, they can be inserted into a cell in a mouse's brain and can be used to turn the neuron on and off, so that it fires in exactly the way you decide. When the laser is turned on, the neuron is also turned on and fires a nerve signal. Using this technology, researchers first isolated the small neuronal network that was activated within the hippocampus of mice as they explored their cage. Then they implanted the light-controlled switch. Next they put the mice in a different cage, and delivered a small electric shock to their feet while simultaneously activating the memory of the first cage. In this way, the mice learned that the memory of the first cage, where there had been no electric shocks, was related to feeling pain in their feet. Back in the original cage, they acted as if they were expecting to get shocked. The researchers could trigger the mouse's fear response and get it to stiffen up in anticipation of an electrical shock, even if they had never been shocked in that particular environment. A false memory of receiving pain had been created.

What is the benefit of manipulating memories like this? Isn't it rather heartless to play with the feelings of small rodents? In a dystopian nightmare scenario, we can imagine that evil super-villains use technology to tamper with people's minds, whether they are enemies of the state or living in a dictatorship. However, this research is not being conducted to gain world domination over rats and mice, but to demonstrate the basic mechanisms behind memory in the brain, on a cellular level. Perhaps in the future we will be able to physically do something to weaken extremely painful memories, or to improve memory in people with memory loss.

Fortunately, nobody has tried to tamper with the human brain to change our memories the way they have with mice and rats. To plant memories in humans, we must turn to psychology. It's been proven beyond doubt that we can manipulate people's memories by pushing the reconstructive nature of memory to its limit.

Elizabeth Loftus and her colleagues have conducted many creative experiments in which they convince their guinea pigs, mostly students, that the most ridiculous things are true. Thanks to Loftus, who is now over seventy, false memory has become a major area of research in psychology. She, her researchers, and her successors have conducted numerous studies in the area. She stumbled into it in the 1970s, after she heard of something resembling an experiment that was conducted on American TV, for entertainment. The network showed a staged crime, and viewers were to call in and vote on who had done it. The staging was very realistic: over the course of thirteen seconds, a man robs a woman, knocks her down, and runs away. As in real robberies, the lighting was bad, everything happened quickly, and there was a lot of movement and many distractions, like screaming and shouting. Crimes like this are complicated, compounded by the fact that witnesses are taken by surprise.

Two minutes later, viewers were presented with a traditional police lineup, containing the robber alongside five innocent volunteers, and were encouraged to call the network to identify the perpetrator. More than two thousand people called in to say who they thought was the guilty person, but the result was astoundingly discouraging. Roughly equal numbers chose each of the people in the lineup, with only 14 percent pointing to the right person. If we take into account that one of the possible answers was that none of the six in the lineup was guilty, the result was the same as if the "witnesses" had guessed blindly. How could so many remember so poorly, even though they had watched the incident with their own eyes? The TV show made psychologist Elizabeth Loftus curious about how false memories are created, inspiring a completely new area of research.

One of the experiments she conducted was intended to make test subjects believe they loved eating asparagus. Before and after the experiment, they monitored subjects' eating habits, and found that after they had planted a false memory about being particularly fond of asparagus as children, the subjects began to buy more asparagus, happily pay more for asparagus, and order it more often at restaurants. The opposite happened when researchers convinced the test subjects that they had once eaten a bad egg. Even those who at first denied that they had been food-poisoned by an egg were, after their meeting with the psychologists, more skeptical of all types of egg dishes and bought fewer eggs. Loftus also tested how different ways of phrasing questions can influence memory. One group of test subjects watched a film where two cars collided. Afterward, they estimated how fast the cars had been driven. Those who were asked "How fast were the cars being driven when they crashed?" thought that the cars were going a lot faster than those who were asked "How fast were the cars being driven when they hit each other?" The wording of the question also influenced how they saw the collision in their memory. Those who were asked about "the crash" saw shards of glass, which weren't there at all.

Loftus also made people believe that as children, they had once been lost at a shopping mall. Her techniques were so convincing that they changed central childhood memories. "I came up with the idea for that experiment while I was driving some friends to the airport, and we drove past a mall. That's how it often works—I get an idea for an experiment just like that," she tells us.

Today, Loftus is ranked as one of the hundred most influential psychologists of the twentieth century, on the same list as Freud, Pavlov, Skinner, and Alexander Luria, who researched Solomon Shereshevsky.

Stories can be very convincing. Memories and stories are strongly connected, as they are in the life story we author throughout life. Perhaps it's the appeal of telling a coherent story that makes witnesses believe in their own guesses after observing a crime? In the experiment shown on TV in 1974, 14 percent recognized the real robber in the confrontation. Hypothetically, this could be due to their exceptional attention and memory, but it could just as well be the result of guessing. When you witness a crime and what you remember can determine if a person is arrested and tried, it may—subconsciously—be tempting to complete the story by offering an answer. When you have given an answer, your assessment continues as a memory, which is almost impossible to separate from the original, volatile memory of the robber. His face was, after all, visible on the TV screen for only three and a half seconds.

In 1844, master storyteller Edgar Allan Poe managed to trick Americans into believing that the first transatlantic crossing in a hot-air balloon had taken place. In bold type in a special edition of the *New York Sun*, he painted an impressive picture of the alleged accomplishment. "ASTOUNDING NEWS! BY EXPRESS VIA NORFOLK: THE ATLANTIC CROSSED IN THREE DAYS! SIGNAL TRIUMPH OF MR. MONCK MASON'S FLYING MACHINE!!! Arrival at Sullivan's Island, near Charlestown, S.C., of Mr. Mason, Mr. Robert Holland, Mr. Henson, Mr. Harrison Ainsworth, and four others, in the STEERING BALLOON 'VICTORIA,' AFTER A PASSAGE OF SEVENTY-FIVE HOURS FROM LAND TO LAND. FULL PARTICULARS OF THE VOYAGE!!!"

Poe included details that could have been true. He used—or perhaps misused—names of people the general public already knew and drew a convincing map of the craft and the route. For Poe, playing with fact and fiction was joyful—because what's really true, anyway, when you look at the big picture? His way of mixing the two is not unlike the way our memory handles recollections and truth. If memory were a writer, she'd have a lot in common with Poe. But it is one thing to trick all of the American eastern seaboard into believing that *someone else* had crossed the ocean in a hot-air balloon. It is something else to trick people into believing that they themselves had been in the basket.

That's exactly what one of Elizabeth Loftus's former students, Maryanne Garry, wanted her test subjects to believe. Garry and her team showed volunteers several photos from their childhood and asked them to talk about them. One of the photos, however, was photoshopped so that the subjects could see themselves, as children, soaring through the air in a hot-air balloon. Fifty percent of the participants claimed they could remember the hot-air balloon and proceeded to describe their flight in detail, even though they had never been on such a ride. They had made themselves a genuine false memory on the spot.

How do researchers convince their test subjects that something that isn't true has really happened? They prepare themselves well and tend to collude with "the victim's"—that is, the test subject's—family and friends to learn as much as possible about the person they are going to lie to. Presenting a story that includes significant, true details increases the likelihood of its being believed. This was Edgar Allan Poe's technique for deceiving newspaper readers. But simply presenting a false story is not enough. A combination of authority, veracity, and interview techniques is also needed. The researcher conducting the interview is a credible person, perhaps someone with an office filled with psychology books and certificates on the wall. A lab coat helps too. The interviewer should also give the impression that they know details about the event—of course, without divulging them all. It is, after all, the participant's memory that is about to be tested. The researcher offers up *some* details, planting a seed and later watering it with follow-up prodding along the lines of "Try to remember; most people need some time before the memory appears," "It's totally normal that you don't remember very well when you haven't thought about it for so long," "Try to picture it," and—when new details make themselves known—"What happened next?" The whole thing is wrapped up in a blanket of kindness, security, and help- fulness from the interviewer. The interviewer is trained in good interview techniques: close to ten hours of training before conducting the interview is not unusual. How to best present images and other types of material to the test subject is drilled, down to the smallest detail. No details should stand out in an odd way or create suspicion.

Now that we're equipped with the recipe for false mem- ories, the question occurs to us: *Who can we trick?* People like us can plant false memories, can't we? A psychologist and an author can quickly become first-class manipulators, right? The only thing is, it feels very wrong! We can't just haul in the first person we see on the street, or one of our friends! It feels dishonest, and not only that, we would risk both the license to practice psychology and a friendship. All signs point to one person, the subject who most shares our interest in showing the consequences of the constructive, whimsical nature of memory, and the one who has every- thing to gain by taking part in a fantastical hot-air balloon trip: our book's Norwegian editor, Erik.

We can't wait to send him skyward.

But before we begin, we have to ask Elizabeth Loftus some questions. How did she manage to trick so many peo- ple in her research? Does she like practical jokes?

"No, not at all! This is only for science. The research on false memories has huge practical consequences; that is why we do this," she patiently points out over the telephone from Los Angeles, nine time zones away.

We simply have to try this ourselves; we want to see it with our own eyes. It will be our gift to Erik: a happy and completely false memory of soaring over Oslo in a brightly colored hot-air balloon when he was five years old.

We pounce on the task, filled with beginners' courage and spurred on by our knowledge of Loftus's successes. Asparagus! Crashes! Eggs! Erik's wife provides us with vital information about his childhood and a bunch of photos. Then it's time to photoshop. A professional designer patches a photo of a hot-air balloon from the 1970s together with one of Erik, five years old, and smooths it out until it looks like a regular photo. His little face pokes up over the edge of the balloon basket, which is still planted on the ground, about to take off. The dumbfounded look on his face suits the situation. Maybe he is a little nervous about flying in the big balloon? It looks completely convincing. Then we book a meeting with our editor "to discuss the book and conduct an experiment about childhood memories."

In his office, we present him with the pictures, the real childhood photos and the one false one. It is the second to last one in a pile of five. Photos from a birthday, boat trip, school, Erik as Superman—the last a good choice; someone who likes to fly will surely be excited about a hot-air balloon trip.

While we interview him about the three first pictures, which he tells us a lot about, our hearts start beating faster. We are about to lie to someone, for real! But what happens when we lay down the photoshopped image surprises us too.

Our editor looks startled and exclaims, "Is this a manipulated picture? I can't remember doing this! No, this is strange. Where did this photo come from? I don't think it looks like me. What is this?"

We have to calm him down, make him think that he simply forgot the incident, and plant little ideas about what might have happened when the picture was taken and why he doesn't remember it—at all! For people like us who aren't used to lying, the urge to break down and confess is overwhelming, especially when our editor picks up the picture and gazes at it in open disbelief. But we stick to our guns. Maryanne Garry, we remember, managed to extract false hot-air balloon memories from half of her unsuspecting participants. It's still possible to make this balloon rise.

"We approached several sources to get these. There are some photos you know from before and some you haven't seen for a while," says Ylva, the psychologist, calmly.

She runs the show now, because her authority as a specialist is one of the few advantages we've got. She tries to wrap Erik in a blanket of calming expert talk—smiling, not too pushy.

"Just take your time. It is completely normal that you don't remember it. We can't remember everything from our childhoods."

It's all about getting him to accept that the balloon ride may have happened, even if he can't remember it. Most of us have had experiences we can't remember afterward.

"Perhaps it was such a strange experience that you didn't have anywhere to place it in your memory?"

It feels fundamentally wrong to feed him one memory myth after the other. Of *course* he would have remembered a hot-air balloon ride. *A special, outstanding incident is more likely to be preserved by the hippocampus.*

Erik's face is still distorted by skepticism. He is desperately scrambling for the truth—is this a fake or a real photo? At the same time, we search his face for some tiny sign that he is slipping.

"Haven't you ever talked about this event at home?"

Erik shakes his head, but he seems calmer now. We have made it past his shock, luckily. He is prepared to negotiate.

"When you were a boy, you were fascinated with the idea of flying. We were talking about the picture of you in the Superman outfit a moment ago. Maybe it helps if you imagine yourself flying? Maybe that will trigger some memories?"

"I think I can picture myself standing in a basket like that, that might have been attached to a hot-air balloon; that I can feel." Erik tries to help. He is, even in this unsettling scenario, a pleasant and courteous man.

"If you can imagine it, it must mean that you've had the experience; all imaginations come from somewhere. Just try to think about going for a ride in a hot-air balloon, and it will come back to you," Ylva the specialist assures him—completely the opposite of what we know about memory.

Because, no, imagining something definitely does not mean that it's true. If so, novelists everywhere would be living their daily lives with some pretty remarkable truths. Imagining something just means that it can, with some distortion, *seem* true. At that point, we're halfway toward a false memory. Erik is an author himself and has a vivid imagination—but does he want to come with us up in the air? In our minds, he is already far beyond the treetops, over the fields. He can look out across the Oslo Fjord and see his sisters below, patiently waiting for their turn. His father is with him in the basket, his arm around the boy's tense little body, which, in a mix of adrenaline and joy, finally gets to fly.

We tell Erik to keep our conversation a secret, another trick we have learned from the researchers. Isolate the test subject, let him brood further on the fantasy of what might have happened, and perhaps, just perhaps, the cuckoo egg will hatch and let the bedraggled baby monster emerge so it can grow into a big and strong hot-air balloon ride in his imagination. Without aunts or uncles or old friends offering the correction "No, you've never been in a hot-air balloon," the fantasy, the dream of soaring, might lift the memory off the ground.

We leave the meeting with mixed feelings. As soon as we're out of earshot, we let loose with laughter. "Do you think he bought it? Good Lord, I hardly managed to keep a straight face!"

There was no hot-air balloon ride during this first meeting, but we haven't given up. This was only the beginning. We didn't expect it all to happen at once. This process takes time, up to three cautiously conducted interviews, combined with the subject's brooding alone over the idea and how likely it was to have happened. The false memory will slowly sink into the network of memories and get caught in the fishnet.

When we arrive the second time to talk to Erik about the false memory, we bring some questionnaires, to make it seem more scientific, and plan to hypnotize him, which should relax him and make him more willing to get into that balloon basket.

But when we take out the file with the pictures, he suddenly says: "I've thought about this since the last time we saw each other. You're trying to trick me, aren't you? This is a fake photo, isn't it?"

For a nanosecond, we get a stabbing feeling in our stomachs and reevaluate whether we could keep pressing, just a *little* bit further. But no, all we can do is put our cards on the table. Perhaps it was due to the fact that our meeting happened on April Fool's Day, perhaps it was just the guilty looks on our faces, but this balloon would never fly. Maybe we stretched the tethers too much during our first meeting, maybe the ropes began to tear as a small, fragile flame breathed the first puffs of air into the red and yellow fabric.

Now, as we sit in our editor's office, our balloon has deflated and lies, flat and airless, across the floor. All of us are laughing at our unsuccessful attempt at memory manipulation, which, in hindsight, we realize was bound to fail. Creating a false memory is hard work, and our efforts were not thorough enough.

The problem with using our editor as a subject is that we, his authors, don't hold the authority in our relationship with him, even though we tried pulling the psychologist card. At best, we're equals. And Erik should also, as a good editor—which he is—know what's in our book and remember that we are writing a chapter on false memory. Even worse is that he, in his position of power, is supposed to be critical of most of what we show him—even pictures. In other words, we'd made our job particularly difficult for ourselves. And it's not just that it's harder to trick an editor than a student, for whom a professor is an authority. It has also been sixteen years since the original balloon experiment was conducted, and in that time people have learned about photoshopping. Today, everybody is aware that photos can be convincingly manipulated.

We aren't alone in our failure to plant a false memory. The researchers Chris Brewin and Bernice Andrews are skeptical that it's as easy as Elizabeth Loftus claims. They carefully examined a large number of experiments on false memory that were conducted all around the world. They found that many of the experiments may not have been measuring false memories at all. Some asked participants to answer how *likely* it was that a memory was genuine on a scale from one to five. The people who experienced a memory manipulation rated the memory more likely to have happened than those who didn't have their memories manipulated. But this is not the same as saying that the participants had actually constructed a new memory. A *genuine* false memory—if we can call it that—requires careful preparation. In more complex experiments where memories are manipulated through multiple interview sessions, false memories seem to appear, while simpler experiments, where the subject's only task is to imagine how things *could have been*, don't show any conclusive effects. Based on this, Brewin and Andrews believe that we have to rethink what false memories really are, and how often they really happen. Yet others consider Brewin and Andrews's work controversial and believe it's an attempt to downplay the very real effects false memories can have on people's lives. It's a thorny subject.

It's fascinating that it is so difficult to trick someone into a false memory when people have created their own memories of unbelievable things with zero outside influence. Just look at the False Memory Archive and Svein Magnussen's tall tale about the car. When they sprout up in the wild, without excessive pruning or outside intervention, false memories grow better—likely because they entwine themselves into our life story. In natural surroundings, they are more convincing.

Maybe we should have tried to trick our children instead. They believe everything we say—or do they?

"There is no difference between adults and children when it comes to false memories. There have been examples of children who absolutely refuse to agree to things that aren't true in police hearings, which demands great mental strength. So children won't allow themselves to be led to false memories any more easily. The percentage of children who experience false memories, compared to adults, is not larger," Svein Magnussen tells us.

Why, then, has it been so important for researchers

to make people believe in false memories? Why is it so important for Loftus to trick people into liking asparagus?

The point is not the asparagus or the eggs. Loftus's research has saved lives and changed the whole justice system's relationship to eyewitness testimony. When she began her colorful experiments in the 1970s, everybody was convinced that what independent witnesses said they saw was the truth—well, it had to be the truth! Why shouldn't it be? And when people confessed to a crime, after weeks of interrogation, it was because they were guilty, weren't they? Back then, the criminal justice system relied on the idea that memory functioned as a precise documentary: expose the film, and you'll have a murderer. But as we've said already, memory is not like that; it is reconstructive.

"Our memories are not made for the criminal justice system," Anders Fjell of the University of Oslo says. "They aren't set up to help us to remember details like the color of a bank robber's clothes. When we are scared for our lives, there are other things on our minds. Memory is an important tool that helps us avoid danger, and in that way, it functions well."

Loftus has examined what happens when test subjects tell a detailed story about something they have observed and are then presented with a written report of the same event in which some details have been changed, such as a jacket's being brown, not green. Many don't react to the planted mistake and begin to believe it to be true. This means that central evidence in a criminal case can change when, for example, a detective listening to a witness's observations makes a sloppy transcription for the witness to approve. Without the witness's noticing it, a typographical error can become a false memory, which in turn can lead to the wrong person being convicted.

"The problem is that false memories are really similar to 'true' memories; even the emotional strength is the same. Most of the time, it doesn't matter whether I had a pizza or a burger last night, but in a criminal court case, such details may be detrimental. We will always have to rely on eyewitness testimony in criminal court cases; we will never be able to rule out the importance of someone's memory to solve cases. All we can do is improve our knowledge of how memory works, in order to minimize the risks," Loftus says.

When the Innocence Project (a nonprofit dedicated to exonerating wrongfully convicted people) examined three hundred cases where DNA evidence has later acquitted the convicted person, it turned out that in three-fourths of the cases, a witness had pointed out the wrong person and gotten that person convicted. In each case, the witness had been absolutely sure that he or she saw the innocent person run away from the scene of the crime or lean over the victim with a gun—or similar observations with great judicial consequences. These well-meaning witnesses had nothing to gain by pointing to the wrong offender. It's just that memories are fallible.

Svein Magnussen has been an assisting expert witness in several criminal cases and has prepared a curriculum for Norwegian judges and lawyers. He has written the standard work on eyewitness psychology in Norway. He knows of at least two murder cases in Norway where there were no corpses or even missing persons, likely caused by someone's false memories of having witnessed a murder, and claims of group sexual assault as part of satanic cult rituals, which have later been proven impossible.

How is it possible to erase the difference between imagination and memory? How can false become true?

"We just don't know," Magnussen says. "I believe that most of us carry around false memories. We think we remember things we have not experienced. But they are often insignificant things. We don't notice them because they're not that important. Even Freud described false memories, but of course they had no significance on the therapist's couch. In the courtroom, though, they have great consequences."

Since publishing his book about eyewitness psychology and false memories, Magnussen has had several inquiries from people in therapy who have been helped to "remember" abuse.

"False memories can result in real traumas. In these people, the false memories are ruining their entire personal history and their relationship to people around them."

Memories of serious traumas seldom appear out of the blue after having been forgotten for many years, despite what you've heard about repressed memories. Most people exposed to serious abuse don't walk around blissfully ignorant until they suddenly remember the horror. A study of 175 men and women who had demonstrably been exposed to abuse as children showed that those who had been abused always knew what had happened. They didn't suddenly recall it in therapy. When asked if they had ever not remembered what they had gone through, they said no.

"To suddenly discover something like that as an adult is close to impossible," Magnussen says.

On the other hand, it may happen under the influence of a therapist who assumes authority and leads their client through the door between the world of imagination and the world of memories.

This doesn't mean that if someone you know all of a sudden tells a story of abuse from their childhood, you can dismiss it as "false memories." Repression can take other forms. It may be that they didn't want to admit to themselves or others what happened but always knew deep inside. Or they have rewritten their life script so dramatically that the abuse wound up in a chapter that they don't reread very often, in order to maintain a necessary relationship with the abuser, whom they may depend on.

Adrian Pracon has a striking false memory from his traumatic experience on Utøya. The last thing he saw before he was shot was a girl being shot and dying by his side. For a long time, he thought that it was one of his best friends he saw dying. It wasn't until he was writing his book that his version of events was overturned. A girl did die by his side, but not the one he thought. His friend was shot on another part of the island.

"I was completely certain it was her, but the girl who actually was killed beside me didn't even look like her. Not even the color of her hair. The only thing they had in common was gender. When I realized my mistake, I had to reexamine all my memories from that day, from every single minute of it. What if I couldn't trust anything I had seen?" Adrian says today.

During the course of the trial, it turned out that most of Adrian's memories actually could be trusted, when compared to the other evidence. It was only this one crucial thing he remembered incorrectly.

How could he have mistaken something so important?

"He must have thought of her for a millisecond, and suddenly she was the one being shot. Perhaps he was afraid that it was his friend who was killed. That's all it takes. The thought becomes real and transforms into a memory as strong as a genuine memory," trauma researcher Ines Blix tells us.

It's one thing for the victim of a crime to remember what happened incorrectly. It's something else entirely for a perpetrator. Is it really possible to believe and admit that we've done something we haven't done?

"Nothing surprises me anymore," Loftus says now, after years of conducting experiments on false memories and monitoring the whole field of research.

"In one of our latest studies, we tried to make people admit to crimes they didn't commit, after depriving them of sleep. It turns out that people's memories are more malleable when they lack sleep," she says.

Even without sleep deprivation, it's incredible what people are willing to admit to. Canadian researchers Julia Shaw and Stephen Porter demonstrated this in a sensational experiment in which they managed to convince 70 percent of their volunteers that they had, when young, committed crimes like burglary and armed assault, without any basis in reality. How is this possible? You wouldn't agree on the spot to having done something that serious in your youth, would you? The participants in the study were convinced. They described the incidents in detail, and what they told was just as evocative as a real memory. The result surprised the researchers too. They had originally planned to conduct the experiment with seventy participants but stopped after sixty, since they already had enough false memories for their results to be statistically significant.

How did this happen? As in similar studies, Shaw and Porter started with a selection of both genuine and false stories from the participants' youth. They worked together with the participants' parents, who gave detailed descriptions of real events and their children's adolescence: where they lived at the time of the alleged criminal act and the names of their best friends. They snuck this information into the first description of the false event to give the story a certain hint of truth as well something for the participants to latch on to.

What Loftus and the other researchers have proven is crucial to the application of law in our society. Both what we remember as witnesses and what we admit to as suspects is no longer seen as inherently true and credible. Without Loftus, more innocent people would have been convicted. Judicial systems have changed fundamentally in this respect, and more and more often psychologists are brought into the courtroom to explain to the jury what a false memory is. The American Supreme Court has recognized what false memories are and knows that the general population from which jury members are chosen still doesn't know how contorted and accidentally changed memories can be.

"When I started working with false memory in the seventies, it was because I wanted to work with something that has practical consequences for people. It has of course been personally rewarding to contribute to a safer legal system in the US," Elizabeth Loftus says.

In the United States, a confession-focused interrogation culture prevails. Police officers have sometimes used rough and unpleasant methods to solicit confessions from people who couldn't possibly have done what they were accused of. For example, a woman was raped in Central Park, New York, in April 1989 and a total of five young men confessed to taking part, even describing in detail what had happened. DNA evidence later revealed, however, that none of them could have been the guilty one.

The Norwegian Centre for Human Rights is located a stone's throw from the new National Museum, overlooking the Oslo Fjord. At the Centre, there are no memory researchers. Nobody here spends days repeating meaningless words in order to count how many were remembered, in or out of the water. Those who work here are lawyers and political scientists—and one police officer. The latter now greets us in a small seminar room on the second floor. How did a homicide detective end up here?

Everything started with the Birgitte Tengs case. Detective Asbjørn Rachlew has delivered hundreds of lectures about his work, and every time he has had to talk about Birgitte Tengs's murder and the ensuing miscarriage of justice involving her cousin. It is one of Norway's most cited criminal cases, but it is also part of Rachlew's personal memories, his life script. That case changed the direction of his entire life story.

On a late May evening in 1995, the life of a cheerful seventeen-year-old girl from a small city in southwestern Norway, till then known for its beautiful heather moors and sandy beaches, was brutally cut short. A few years later, the interrogation methods used during the investigation of her homicide made history. Tengs was the image of Norwegian innocence. A photo of her dressed in folk costume with blond, wavy hair cascading over her shoulders was shown in newscasts. The police were faced with a murder by an unknown offender in a small, safe island community. When Tengs's cousin ended up in the police spotlight, they recklessly pushed for a quick resolution. There was just one problem: the cousin denied committing the murder. The interrogators were determined to break him.

In gangster films and crime TV shows, we have all witnessed the good cop/bad cop routine. The methods used when Asbjørn Rachlew started out as a detective were very similar. A tough tone, the "we know you did it; it's just a matter of time until we have the evidence" tirade. They conducted long interviews without breaks, and the accused was seated on a lower chair than the detective, so he had to look up at him. The chair was old and rickety, set up to make him realize that this was a place where people remained sitting for a long time, so there was no use fighting it. The suspect had contact with only one detective, to create a kind of dependency, and the detective had the sole power to stop the interrogation, or get water, food, coffee, tissues. The detective was supposed to, little by little, become a person to confide in. He could also tactically use body contact to create trust—put his hand on the suspect's shoulder, his arm.

"We learned the interrogation techniques from other, more experienced detectives. These were not standardized methods; each detective had their own signature moves. We copied American interrogation methods and were inspired by whatever could give us some guidance regarding the questioning—films and TV series," Rachlew tells us.

What happened when the police located a suspect—Tengs's own cousin? The detectives isolated him. Two of them took turns having sole contact with him, and they forced him through long interrogation sessions. They told him that he most likely had repressed the memory of the murder because it was just too horrible to think about. They had methods to help him bring back the memories. He was supposed to imagine how it hypothetically could have happened, had he been the murderer.

Does it sound familiar, like the hot-air balloon experiment we conducted on our editor? It is not a coincidence. The methods used by memory researchers trying to create false memories in test subjects come directly from the police interrogation room.

His isolation and desperate situation wore on Tengs's cousin. He was promised rewards if he cooperated. He eventually began to look at his time with the detectives as welcome social contact. He was caught in a bubble, and the detectives made it shrink. After a while, his reality was just the one he shared with them, in the interrogation room. He cooperated. He implemented the thought experiments, imagined the gruesome acts. He began to doubt his own understanding of reality. Could he really have repressed it? What he thought he remembered of that evening, the trip home from the city, was it just constructed afterward to cover up this unthinkable evil? The tasks the detectives gave him became more and more concrete. He had to write stories about the murder. He penned several versions, imagining the evening it happened and possible scenarios for how it went down. He had grown up in the same town as Tengs. He knew the area well and could clearly see the heath, the heather, the path where her body was found. He felt the adrenaline in the tense situation—we can all imagine the feeling. When he wrote a well-executed story, he was rewarded with goodies and social contact with the detectives. The tissue box was pushed across the table toward him, the chair was moved closer to his, a hand was gently placed on his shoulder; someone understood. His written story gradually became more and more similar to the story the police had managed to reconstruct from other evidence. Tengs's cousin was convinced of and confessed to the crime—but DNA found at the crime scene belonged to others.

There are three different types of false confessions. There are the voluntary ones, when people confess to something they haven't done because they want the attention or because they think they deserve blame in a broader sense (perhaps they're guilty in the eyes of God—who knows). Then there are the forced false confessions. People may make these after torture or long periods of being pressured. Many suspects think it's easier to confess to escape immediate stress or pain, hoping that the judicial system will clear them later on. They are often wrong. Their earlier confession will accompany them and appear as proof of guilt. The third type is when they themselves believe they've done it and freely make their own confession, based on a false memory. Which type of false confession the cousin made, no one is completely sure. It is possible that, right then and there in the bubble, he briefly owned the evil story about himself as a rapist and murderer. Even without an episodic memory of the event, it's possible that he believed he did it. In any case, his confession was immediately rewarded—of course, not in the form of release, but in the form of resumed contact with the outside world. It didn't take long for him to withdraw his confession and assert his innocence.

Tengs's cousin was convicted in criminal court and later acquitted in the court of appeal. But in a civil suit initiated by Birgitte Tengs's family, he was sentenced to restitution. In civil cases, guilt is determined based on a balance of probabilities, whereas in criminal cases, guilt has to be proven beyond a reasonable doubt. The cousin sued the Norwegian government in the European Court of Human Rights for wrongful prosecution and won. And Tengs's true murderer is still on the loose.

In the aftermath of the case, the Norwegian police were raked over the coals by the Icelandic eyewitness psychology expert Gisli Gudjonsson, who dissected the interrogation methods that had been used and showed how the cousin could have been manipulated into producing false memories. He also referenced research literature on eyewitness psychology, including the work of Elizabeth Loftus.

As the ripples from the Birgitte Tengs case spread far beyond the small island community where she was found murdered, Rachlew was working as a homicide detective with the Oslo police. When it was revealed that the cousin's confession had been solicited under pressure, the media also started to investigate. What methods had the police actually used? What had gone wrong?

"Of course, we shrugged collectively when the media interviewed Gisli Gudjonsson and British interrogation experts. They spoke of totally different methods than those we were used to. But it made me curious. Suddenly I understood that there was a whole discipline I didn't know about. Why hadn't I heard of this before?" Rachlew says today.

His job at that time involved conducting interrogations of murderers and gang criminals with the intent to manipulate them into a confession. When approaching the point of a possible confession, the detective would showily put out a box of tissues to appeal to the suspect's feelings, and lean in with a sympathetic touch on the arm.

"Physical touch to push the suspect over the edge and finally help him ease his conscience was something we used more and more systematically," he tells us.

It was outright manipulation; he knows that too well today. He even developed his own signature maneuver, which, together with his charisma, made him a high-achieving detective early in his career.

"I am not proud of this. What I am most ashamed of is how I screamed at a suspect that I'd be back and he wouldn't know when."

Sitting by Rachlew's side this afternoon is his daughter. She is writing a letter with the deliberate lettering of a five-year-old. It's hard to imagine this calm man, who is patiently and quietly helping his daughter spell a difficult word, shouting at suspects in murder cases.

The pressure surrounding the Tengs case forced Rachlew to make a choice that would change the Norwegian police and judicial system forever. He took time off work to study British interrogation methods in the Investigative and Forensic Psychology program at the University of Liverpool. Today in the United Kingdom, the police use an interview technique called "investigative interview." Contrary to the confession-driven interview form of the old days, where the goal was to get a confession at whatever cost, the purpose of this interview form is to ease out as much information as possible from witnesses' and suspects' memory, whether or not they are thought to be guilty. You are supposed to secure important evidence from both suspects and witnesses.

"Our methods were totally unscrupulous!" Rachlew understood, shocked, in the course of his education in England.

In the work for his PhD, he investigated a case that had affected him deeply: a rape that had taken place on the outskirts of Oslo. The woman who was raped was also the only witness. How could they secure evidence?

The woman gave a description of the rapist: "Male, approximately forty-five years old, approximately six feet tall, a slightly stocky built, a bit of a paunch. Short, dark-brown hair, graying, thinning in front. Very bad teeth, possibly some missing. Medium-dark complexion, possibly southern European, perhaps Turkish. Spoke broken Norwegian and said that he had lived in Norway for ten years."

A facial composite was sketched and distributed to national media. Tips came in to the police. The drawing looked like a Bosnian-Norwegian family man who denied any involvement in the rape. The woman saw a picture of him in a photo lineup. He did look like the drawing; it must have been him. But she wasn't completely certain yet, so a live lineup of people was arranged. The suspect was presented along with six other men, all police officers or interpreters, all with similar features. All of them knew they were safe from prosecution except the Bosnian, who was so nervous he couldn't follow the instructions of keeping his arms by his side and standing to show his side profile. He clearly resembled his picture from the photo lineup. The photo looked like the composite sketch. The composite sketch looked like the woman's memory of the rapist. And the memory…? It's like a memory telephone game. The memory of the rapist gradually became the memory of the suspect. It doesn't mean the woman's memory was poor, or that the traumatic experience had weakened her judgment in any way. This is just the way memory works. It is a living organism, always absorbing images, and when new elements are added, they are sewn into the original memory as seamlessly as only our imagination can do. Furthermore, all of us have different abilities to remember faces. There is actually a special part of the brain set aside for just that; that's how important it is to us. Some of us need to see people several times before we are able to recognize them again the next time. It's likely that a rape victim will remember the face of a rapist clearly. But even if the memory of a traumatic incident is stored well in the brain, that memory isn't spared the art of reconstruction when we try to remember it again. We know that from Adrian Pracon's story.

The Bosnian family man was first convicted in district court. This happened despite the fact that there was DNA evidence, from someone who wasn't him, on a pair of men's briefs found at the scene. The witness testimony was given the most weight in the trial, and she was certain that he was the offender. But in appeal court, he was acquitted.

What made the case so heartbreaking was how it was finally solved. The DNA from the real rapist later appeared in the DNA registry as part of the evidence from a serious new case: he had killed his wife.

"A woman lost her life, and a child lost both parents when he became a murderer, and we could have prevented that," Asbjørn Rachlew says.

After Rachlew's return to Norway after his studies, the reforms he instituted were hard on all involved. For some of his colleagues, the transition to the new British methods was difficult. Many of them didn't speak to him for several years and avoided him in the corridors. But today he has resumed contact with many, and most now understand why the old rules had to go.

Rachlew's time in Liverpool spurred the beginning of a new era in Norwegian policing. Based on the British model, Rachlew created guidelines for a Norwegian adaptation of the investigative interview. Today this method is used by all Norwegian criminal detectives, and their training includes, among other things, basic instruction in how short-term and long-term memory work, and how false memories come about.

"The consequence of not knowing how memories work is that detectives don't treat testimonies from witnesses with the same strict discipline as they do with hard physical evidence," he explains.

Picture a murder scene on your typical TV crime series. The crime scene is teeming with technicians in white or blue overalls and face masks. They cautiously move around, picking up evidence with tweezers and putting it in marked bags to be sent for advanced analysis. Now Rachlew shows us a picture of a beautiful forest path, covered with a fresh, velvety carpet of new snow. In the forefront, the path is cordoned off with police tape.

"The evidence lies under the new snow. You can't just randomly stomp around. Every step you take leaves a track that changes the truth. It's the same with witness evidence," he says.

Most of our common-sense beliefs about witnesses, confessions, and detective work come from books and TV. And most of how memory is represented in the crime genre is flawed.

"Crime literature has definitely contributed to creating a false idea of how memory works. For example, the truth in witnesses' statements is seldom questioned," Jørn Lier Horst says. He was once a top Norwegian police investigator and is now a best-selling crime writer. His books have been translated into a number of languages, including English and Japanese.

In April of 2016, he was given a Polish award for realistic presentation of police work in his books.

Eyewitness psychology plays an important role in the plot of his crime novel *Ordeal*, in which—without revealing too much—a false memory is gradually uncovered. Those familiar with the literature about eyewitness psychology can find clues throughout the book pointing toward a false memory. Even the investigative interview method has its place in Horst's books; he has taken a course taught by Rachlew.

"The interrogation methods used in crime films and books are totally wrong. Behaving boorishly in front of witnesses, interrupting them in the middle of a sentence or threatening them; this is not the way the police want to work to get at the truth. Often, in these stories, they make a big mistake which we would never make in real life. We never interview two witnesses at the same time. You can see that on television: a married couple is summoned for an interview, and the detective talks to them together. That would never happen. They're supposed to give one statement each, independently, of course," Horst says.

He thinks many crime witnesses are surprised by how little they remember when they compare it to what they've seen in crime stories.

"Witnesses are unreliable, and they often remember things incorrectly. As a new policeman, I once worked on a much-publicized murder case where four witnesses described seeing the suspect riding a moped. The problem was that the descriptions were all totally different. A couple of the witnesses even remembered the license plate on the moped. It turned out that the color-blind witness had the best description of the man who was later convicted," Horst says.

How would the eyewitness psychologist's *CSI* look? Well, for example, they wouldn't make the mistakes the Stockholm police did when the Swedish foreign minister Anna Lindh was stabbed to death in a department store in 2003. The witnesses were immediately ushered into a back room, where they sat together, waiting for the police to question them. Five of those witnesses said the suspect had been wearing a military jacket. This description was sent out to the media and all airports and border crossings. When the police later watched the surveillance footage, it turned out there was no military jacket to be seen anywhere. What happened was that one of the witnesses had tampered with her own memory—of course not deliberately—and had "seen" a military jacket, probably because she associated it with violence. She had then talked about it with other witnesses in that back room, and her mistake had spread.

Now when witnesses are called in, the police work *with* the fundamental laws of memory, instead of against them. They try to avoid helping mistakes sneak into a witness's testimony. First, they establish a good relationship with the witness. Memory works very poorly under pressure.

"Just think about how difficult it is to find your wallet when you're dashing around your apartment, stressed because you can't find it!" Rachlew points out.

Rachlew emphasizes that detectives' openness and integrity are prerequisites for a good relationship with witnesses and make them less anxious. When witnesses know how the interview will proceed, they relax more. They have to be able to trust the detective to be prepared to open up. Then they do the most important job by themselves: describe what happened. Most of the events should, if possible, be retold in the witness's own words, relating their own experiences, without outside involvement.

No leading questions, no manipulative use of tissue boxes. There are prescheduled pauses, there is a predictable framework. When the witness can't think of anything else, the detective can use memory-promoting techniques. The most important of them is based on what we know from, among other research, the experiment with the divers:

being in the same environment as when you first experienced something awakens more memories. Witnesses are asked to picture themselves at the scene and in the situation. More details may then come forward.

"We check what the news and the weather was on the day in question. Perhaps something happened that can help the witness remember the day—such as a skier winning a gold medal or unusual weather."

The follow-up questions need to be phrased carefully so that no new information is eased into the memory of the witness. Everything is recorded on film, so that others are able to scrutinize it later. The method is standardized and totally transparent.

The reason that Asbjørn Rachlew, though he once behaved as a manipulative bully, sits here today in an office at the Norwegian Centre for Human Rights is that he chose not to defend his actions. Instead, he regarded the criticism that was directed at the police (and by extension, himself) with curiosity. He may have perfected the methods he used, but they didn't stand up under scrutiny, so they had to be replaced by something more scientific. Loftus's research from the 1970s may not have reached the Norwegian police until the 1990s, but today Norway is on the forefront of human-rights-friendly investigation methods, mostly thanks to Asbjørn Rachlew and Svein Magnussen.

Reforming interrogation techniques helps us preserve human rights and prevents us from convicting innocent people. But it also ensures that murderers aren't acquitted because important witness information has been lost.

For several years, Rachlew and his good friend and colleague Ole Jakob Øglænd did a lecture tour where, alongside the innocent suspect Stein Inge Johannessen, they described what went wrong in a murder investigation and how their mistakes could have been prevented. The ex-junkie Johannessen, who was the main suspect, was heavily pressured in accordance with the old methods and kept in custody for nine months. A faulty eyewitness testimony was one of the reasons the police strongly maintained their suspicion. But a few days before the case was to appear in court, the actual murderer turned himself in. The lecture tour allowed the three to teach people how the investigation should have been conducted. Johannessen has since passed away, but Rachlew still lectures in schools, and to journalists and investigators, about human rights and interrogation methods, including the investigative interview. He was also an expert advisor during the questioning of Anders Behring Breivik.

Recently, Rachlew traveled to Geneva to give perhaps the most important lecture of his professional career. He spoke at the UN about his methods, how they were implemented, and why they are important. Everything he said was simultaneously interpreted so that those who didn't know English could follow his lecture in their own language through headphones. And so his message about memory functions, human rights, and witness psychology—and the truth—spread across the globe.

Although, the truth. How can we trust our memories to provide the truth? Memory is, as we have said many times now, reconstructive, and there are mistakes and flaws in all our recollections. The difference between a real memory and a false memory is not that the false memory contains mistakes, because all memories are fallible, but rather how wrong they are. Picture yourself putting on a Beatles record and hearing "Yesterday" sung by the Rolling Stones—or the Beatles singing "Satisfaction" instead. A false memory means that the wrong track has snuck onto your record.

"We can never know with hundred percent certainty that a memory is true or false, unless there is objective evidence to support it. We just have to live with this uncertainty, making sure that work is done to facilitate memory as much as possible, avoiding mistakes that we know may mislead memory. Court trials will always rely on eyewitness memory. The goal is to make sure that memory comes as close to the truth as possible," Elizabeth Loftus says.

THE BIG TAXI EXPERIMENT
AND A RATHER EXTRAORDINARY
GAME OF CHESS

Or: How good can your memory get?

"You appear to be astonished," he said, smiling
at my expression of surprise. "Now that I do know it
I shall do my best to forget it."
"To forget it!"
"You see," he explained, "I consider that a man's brain
originally is like a little empty attic, and you have to stock
it with such furniture as you choose."
ARTHUR CONAN DOYLE, *A Study in Scarlet*

S YOU APPROACH London by air, you see the city
sprawled out far below. The Thames shimmers like a
ribbon thrown on the floor by an impatient child.
Two thousand years of history shines back at you—from
50 CE until today, the city has grown without any coherent

plan. It is a chaotic wilderness of streets and landmarks from every century, churches and steeples, prisons and palaces, hospitals and museums all scrambled together; five-hundred-year-old pubs with crooked floors and modern buildings with facades of glass and steel sit side by side. Originally, London consisted of several separate villages that gradually grew into each other, and so the city has no single center but rather many different ones. The streets take strange turns, suddenly stopping or transforming into alleys. It's no coincidence that so many action movies set in London feature car chases ending on foot, with the hero jumping fences and turning tight corners. The heart of Britain is a tight nest of streets with no structure, an untidy assortment of architectonic candy—a city planner's nightmare. How is it even possible to learn the streets of London by heart?

Eleanor Maguire, whom we met in earlier chapters, became famous for her research on London taxi drivers. She showed that there are visible differences between their brains and the brains of those who never acquired the Knowledge, the test all London black cab drivers must pass before they're allowed behind the wheel. The Knowledge requires them to learn the names and locations of 25,000 streets, 320 specific routes, and countless landmarks, all without the help of a map or GPS. The revelation that this visibly changes the brains of taxi drivers may have led the general public to view them in a different light. Who knew that, behind the wheel, they harbored such a stunning neuroscientific secret?

"To our surprise, we found an increase in the posterior part of the hippocampus in the taxi drivers," Maguire says.

London's convoluted maze of streets and alleys from throughout history works brilliantly as an environment for extreme training of spatial memory. While Maguire sat in her office, wondering whom she could research, thousands of the most highly trained memories in the city passed by her window in their shiny black cabs. Suddenly, the idea hit her. Had it not been for London's messy structure and the rigid test the cab drivers must endure, Maguire would not have become known as a leading memory researcher, not only among her peers worldwide, but also among an entire fleet of drivers.

How did cab drivers become the key to some of memory's most tightly guarded mysteries? British taxi drivers spend years training before they take the test and graduate. Many work full-time in other professions so they can live in London, spending all their free time learning routes. You can see them driving around the city marked with an L for Learner and bearing a map clipped to the handlebars of their mopeds, attentively studying each street they travel.

"There were four hundred routes when I took the Knowledge seventeen years ago," a taxi driver named Judy tells us with obvious pride in her voice, accompanied by a certain amount of contempt for those who get away so easily with the test today. She spent two years and ten months studying for the Knowledge, and passing it was a big struggle.

"I am one of few female drivers, so it was important for me to pass on those grounds alone," she says. She refused to give up, even though the instructors did everything possible to stress out the taxi novices during the tests. If you're going to make it as a taxi driver in one of the world's largest cities, you've got to be able to handle some stress.

As a cab driver, Judy has to know how to navigate all the roundabouts and traffic lights in the city, but she also has to be able to drop off customers with the correct address to their left-hand side, so they don't have to cross the street. At the same time, she keeps up with the names of pubs as they change owners, the new hot spots that constantly pop up, and traffic announcements about closed streets or traffic jams. If a London cab makes a detour, it may be because the driver knows that the obvious, quickest way is not so quick on that particular day.

We have now given Judy a difficult task, we think. We're in Bloomsbury and want to go to Short Street, a street only two blocks long located on the South Bank, two and a half miles away. A short, insignificant road is of course more difficult to remember than a major thoroughfare. At first she thinks we're asking for Shorter Street by the Tower of London. But then she gets it.

"Wait a minute. Yes, I know," she says and describes a theater, a pub, and a store located along the street leading up to Short Street. "I can clearly picture the street."

We nod and smile and think that she may be wrong as she drives us through the city toward the stubby little street. We're going in the right direction, at least.

"We're supposed to learn so-called points of interest, and that's how I navigate. If I want to know where to go, I visualize the street. I see it in front of me," she says. She searches her memory for relevant points of interest close to her destination, and when she finds them, it's like a map of the area unfolds in front of her.

And then we're there—after sailing past the little theater, the pub, and the store, just as she described.

"People think the internet is replacing memory, but that's what they said back when they invented books, once upon a time," Eleanor Maguire says. "We are always going to use our memory. Even if we have GPS, we'll need memory to find our way—for instance, inside large buildings, like a hospital. Anyway, taxi drivers make quicker and better decisions than GPS."

She bemoans her own poor spatial memory. She sometimes has to call for help just a few blocks away from conference centers—and a lack of experience isn't to blame, as she travels almost as much as a head of state might in a year.

Her research team has examined the taxi drivers' brains several times. What grabbed everyone's attention the first time was the first visible proof that our brains may change when we are studying. Maguire's research won her an Ig Nobel Prize "for presenting evidence that the brains of London taxi drivers are more highly developed than those of their fellow citizens." Maguire traveled to Harvard to receive the prize, a small trophy supposedly containing a nanoscopic gold nugget. She held it on her lap during the flight back, only to later drop it on her office floor. The gold nugget was probably devoured by the vacuum cleaner. But the London cab drivers are proud to have proof of their superior brains. Perhaps it has had a slight effect on their professional pride, which was, by all accounts, already quite noticeable.

Maguire's research, however, raised more questions than it actually answered. She showed that the cab drivers, with their intense spatial memory training, had a larger posterior (back part) of their hippocampus. But is this really proof of plasticity, evidence that the brain can actually change? Couldn't it be the case that those with large posterior hippocampi might be the ones cut out to become taxi drivers in the first place? To get a final answer, Maguire and her team conducted a much more involved study. This time, they followed taxi drivers from their first day of training until the day they passed the Knowledge. They measured their brains and memory before training started and after it concluded. This time, they were able to clearly show that the training was what caused the hippocampus to change. Before their training started, the aspiring taxi drivers had exactly the same sized hippocampi as most people. Their hippocampi were not visibly larger until after years of zealous effort on their mopeds. This should be proof enough that the brain is trainable. But there was something else: the change only happened in those who passed the test. Evidently, the amount of training someone does affects the result. It could be interpreted as more proof of the training effect. But the reasons for not passing the Knowledge are varied. Some people give up because they can't handle the years of low income and spending all their spare time on training. Some have family obligations that prevent them from training often enough. Another possibility, which is hard to measure, is that those who pass the Knowledge have brains with a greater potential for change than those who don't pass. What determines this—genes, special growth substances in the brain, diet, or something else— we still don't know.

"But what we do know now is that memory adapts to what we need it for, even when we grow older," Maguire says.

Maguire's words offer hope not only to those of us struggling to keep our brains nimble as we age, but also to those whose brains have taken a hit, whether from injury, epilepsy, or even the early stages of dementia.

There is one small problem in specifically training a single part of the brain. Inside the skull's boundaries, there isn't an unlimited possibility for growth, so memory training won't give anybody a gigantic brain. It actually seems as if memory training comes at a cost to other areas, at least in the case of the London drivers.

"The anterior part of the hippocampus—so the front part—was actually a little bit reduced along with the increase in the back," Maguire tells us. The training led to the drivers' performing worse in another area of visual memory. Maguire could measure a statistically significant difference compared to a control group using a simple memory test in which drivers had to redraw a complex figure from memory. "It's as if their brains had to prioritize spatial memory, and this compromised other aspects of memory to some degree," she says.

So how is it possible that the brain can actually change due to experience?

So far, what lies behind the training effects in the brain remains a mystery. When we lift weights, we know for certain that there are changes in the muscle cells because our muscles grow. With the brain, it's not that simple. When Maguire says that she can see a difference in the taxi drivers' brains, she can't, of course, see it from the outside, in the way you can see bulging muscles on someone who has taken up weight lifting. The changes she has seen in the taxi drivers are actually tiny when compared to the whole brain, and can only be detected by comparing groups of trained and untrained people. Nevertheless, there is an actual increase in gray matter. What does this increase consist of? Every single memory we create changes the connections between thousands of neurons, as Terje Lømo demonstrated using rabbit brains. Can these changes in and of themselves lead to a bulging hippocampus?

Long-term potentiation triggers physical changes in brain cells. Cells grow and change so they can deliver a stronger signal. The cell receiving the signal grows more receptors so that it can react more strongly when it receives a message. For a long time, we believed this was all that happened. Up until recently, everyone agreed that our brains were fully developed by our twenties. Our neurons were in place at birth, all of them. *All one hundred and some billion neurons!* From birth until the day we die, they slowly disappear, one after another. Cerebral loss is all we have to look forward to, apparently. But, like the songbirds whose brains can grow in adulthood, adult human brains have been shown to develop new neurons too.

In some animals, scientists have discovered several places in the brain where stem cells "give birth" to new neurons. In the human brain, this has been shown to happen in only two places: the hippocampus and the olfactory bulb (where our brain processes scent stimuli). But we're most interested in these new neurons in the hippocampus. Why would our brains develop new neurons there, if not to grow more room for memory? Memories are stored in several places in the cortex, but it's the hippocampus that has the most important job when it comes to coordinating our various experiences and assembling them into complete memories.

The newborn neurons in the hippocampus have a lot of growing to do before they can become full-grown memory carriers. They must find their place in networks of other neurons, networks that are already connected to other networks in a meaningful way. Otherwise, they end up isolated and lonely, like an astronaut untethered from the spaceship, drifting alone in space. It's been difficult showing how new neurons make that journey. In humans, it is impossible (so far) to follow a neuron as it gradually forms connections to other neurons—its great transition into an active memory trace.

Even in rats' brains this is complicated, but it is easier to demonstrate there than in a human brain. Researchers have managed to measure the activity in newborn neurons while rats attempt to navigate a maze. When new neurons are four weeks old, they begin to synchronize with the neurons that were there before them. This must mean that they are connected to the memory network. Perhaps the new neurons have a special role, acting as a unique signature of a given memory so that it stands out from other, similar memories.

London's taxi drivers have so much professional pride and love their jobs so much that they rarely quit before they have to be carried out of their cabs. But when their adventures in taxi driving are over, they can look forward to a pleasant retirement—and their hippocampus becoming normal again. Gone are the special skills they have spent an entire professional life practicing and maintaining.

"We didn't find enough retired taxi drivers to be able to make real conclusions—they never, ever quit!" Maguire sighs. "But the results do point to a return to a normal state of the brain."

Like rats in an unbelievably challenging maze, London's taxi drivers scurry through their city's labyrinthine streetscape. And exactly like rats, their grid cells and place cells are activated when they locate places in their memory. But so far, nobody has embedded little electrodes deep inside their brains to watch this happen while they drive around, not even May-Britt and Edvard Moser.

"It's not unthinkable that there are grid cells that cover great distances, as in London, but we don't have any methods to study them," Edvard Moser says.

Extreme learning leads to extreme remembering. The brain can even change visibly. But taxi drivers are not the only ones who have to commit something huge to memory. Professional chess players spend their days focusing on chess strategies. Does that mean that chess players also have freakishly good memories?

Simen Agdestein, chess grand master and former instructor of the current world champion, Magnus Carlsen, has spent most of his life playing chess on an international level. In his office, at the Norwegian College of Elite Sport (NTG), he shows us how to prepare for a game of chess. With the help of a statistical program, he can see what typical moves to expect from other players and figure out ways to feint them. It's a matter of memorizing, pregame, what an opponent's weaknesses are.

"In the olden days, we used to study yearbooks filled with descriptions of past chess games," he says, pulling some examples from his shelf. The volumes he flips through with the same nonchalant motion as a magician shuffling a deck of cards aren't exactly summer beach reading. They're just endless lists of dry chess moves. "A57—1.d4 Nf6 2.c4 c5 3.d5 b5."

"I don't remember any of this now, but I used to pore over these books," he says. Now he trains future chess talents.

The Magnus Carlsen room, where the young chess players have their daily sessions, looks like some sort of shrine to the world champion. The walls are covered with his photos, ranging from when he was a little kid up to the championship position he holds as a young man today. We have laid out seven chess boards behind a curtain, partway into a game—somewhere between the twenty-first and twenty-fourth move. Four of them are positioned like well-known games from past tournaments. Magnus Carlsen versus Viswanathan Anand is there—the famous game that made Carlsen the world champion, when he beat the reigning world champion. Three of the board setups are complete nonsense. We have made up the nonsense games by starting from a point in known games, and then switching the position of all the pieces by lottery draws, so they end up completely scrambled. On one of the boards, the two kings are positioned right beside each other, an impossibility in chess. On another, a pawn has inexplicably snuck past the opponent's bishop to stand right behind the enemy line.

The four champions we have challenged get to see the boards for five seconds at a time and then have to reproduce what they saw. Will they remember all the boards equally well? Or will it be more difficult to remember the nonsense setups compared to the logical, realistic, historical games that might already be familiar to the players? We're talking five seconds here. That's the time it takes to unlock your door at home or pour yourself a glass of water!

The boards we've prepared have plenty of pieces, more than twenty on each. How will they manage to remember more than a few pieces? Our experiment, first attempted in the 1940s with Dutch grand masters, was designed to show that chess players have developed an almost intuitive memory of the game. Through endless chess games, they have learned positions, known openings, and typical moves. They can identify nonsense setups immediately, and past games are supposed to remain ingrained in memory. Their chess expertise helps them to absorb and understand what they're looking at more quickly than the average person, and it's easier for them to reproduce what they have seen, even if only for five seconds.

Our first participant is Aryan Tari, one of Norway's greatest chess talents. The shy teenager recently became the world's fourth-youngest grand master and, after our experiment, the Junior World Champion. He fumbles a little bit on the first board, a real historical one, where he can place only six pieces correctly, but then it picks up. His best score is sixteen correct placements in the Anand–Carlsen game. This is a game he has analyzed carefully, as have all Norwegian chess talents, and he knows every move. But when it comes to the nonsense games, Aryan manages at best seven pieces.

In the experiment from the 1940s, the best players correctly placed up to twenty-four pieces. When the experiment was repeated in England in 1973, one grand master remembered, on average, the locations of sixteen pieces in the real games.

Olga Dolzhikova and Simen Agdestein attempt the same task. When Olga places her pieces on the board, her maximum score is also sixteen, but in several instances, she has positioned a black rook where a black queen should be, or a pawn where a knight should be. In her memory there are black and white pieces. She often remembers exactly where each color of piece goes, just not always exactly which piece it is.

"I first look at the middle rows on the board; all the action takes place here," she explains. That's also where she has placed most of her correct pieces. Toward the edges of the board, her memory is more blurry and imprecise.

When chess instructor Simen has his turn, he zooms in on the board in front of him, focused. His gaze darts across the chess board for the five seconds he has available. Very quickly, he grabs the pieces with both hands with the smooth motion of someone painting a large canvas, and swings them across the board. When he steps back and checks his result, only a few pieces are placed incorrectly. He gets twenty of them right. The next board is not as good. He remembers only six correct pieces. This is one of the illogical boards.

"This is nonsense," he says. He frowns and scratches his head with a pawn, looking displeased with himself.

The next time he throws the pieces out randomly. His self-assuredness is completely gone. After this round, though, his successes continue with the real boards. Of course he remembers the game between Carlsen and Anand. He was the expert commentator when it was played and was interviewed about it in the media in his role as Carlsen's former trainer. For a long time, he swings the queen between e1 and d1 and finally lets her land on the wrong one, e1. This is his only mistake on that board. He has twenty-two correct placements out of twenty-three.

"I was thinking that she would be on d1, but then it seemed illogical too. But, of course, there might have been a situation over here earlier," he says, waving his hand over the right side of the board, "which made her go there."

Olga has the highest scores on the nonsense boards, with ten as her best result.

"I remembered it because it was so illogical. Every single illogical position burned itself into my memory. I think it is a memory advantage to be a chess player, even when it comes to the nonsense boards. In my head, everything is chess positions, impossible or possible, and I remember things based on that. I did my PhD thesis in pedagogy, and I found that those who played chess had better short-term memory than others, and I think it is because we are thinking about relationships all the time. We see things in relation to one another," she points out.

For the average person who doesn't play chess, all the boards look more or less the same, but to the professional chess player, the nonsense boards stand out right away. For Simen, the illogical boards were total chaos. When he encountered the first nonsense board, he tried to take a mental snapshot of the board, but that didn't work at all. His next strategy was to try to remember only a few pieces so he could at least get a few right.

For both Olga and Simen, there was a major difference between the nonsense boards and the real chess games in terms of how they appeared in their memory.

"With the proper games, it was like I blacked out for a couple of seconds, remembering nothing. But then an image of the whole board appeared in my mind, like out of a fog. With the nonsense board, there was nothing, no image. All I could do was try to remember as much as possible," Olga says, and Simen nods in agreement.

"But if you tested us again now, on the proper boards, we would do it perfectly!" Simen says. "They are swirling around in my head now. I'll be thinking about them for the rest of the day."

The only one left to go is Jon Ludvig Hammer, one of the world's best active players and Norway's second best. When we test him, the game turns, and we are faced with defeat— the experimenters are suddenly the ones who are tested.

BUT FIRST, LET'S find out a bit more about how much it's possible to remember, and how those with great memories learn the things they need to remember. There are, of course, professions unattainable to those with bad memory.

Actor Marie Blokhus played the main role in *Hamlet* on one of Norway's major stages, Det Norske Teatret (The Norwegian Theater), in a gender-reversed version of Shakespeare's most famous tragedy. The play is heavy with monologue. There is a lot of text to be memorized, long passages that are recited by the main character.

Blokhus remembers the role as being like an ice-blue spiral. Yes, that's how she explains it. She has synesthesia, like Solomon Shereshevsky. Sounds give rise to colors and shapes in her mind. When a trolley filled with props for a play bounces past us, Blokhus says that it makes a brown sound, shaped like a curled-up snake. Sudden sounds are yellow or green; nails on a blackboard make a "funny, pointy, and yellow sound."

Music tends to become complicated geometric shapes, but theater plays and poems also evoke complex shapes and colors. *Hamlet*, as a whole, is an ice-blue spiral to her. That's how she remembers her role.

"Perhaps the blue comes from the sea and the sky and solitude in nature, I don't know. But most of the play relates to that feeling, and then other colors and images superimpose themselves on certain scenes. Blue is the base color. Everything emerges from that. But remembering the role isn't as difficult as remembering to *forget* the role," she says.

Every evening, she stands onstage trying to think that she has never before experienced what's happening. It's supposed to be as new to her as it is to the audience. She doesn't want to deliver a monologue like a robot.

"I have to be completely open to what happens on the stage; otherwise it won't seem real. I expose myself and my overwhelming feeling of loneliness, veiled by the character of Hamlet. The words are part of me. I often pace back and forth when I'm rehearsing, to make the words enter my body. And when I'm onstage, I have to trust that they're there, somewhere, in my body," she says.

Through her actor's training, she has learned several techniques for analyzing texts and characters, but she doesn't use any purely mnemonic techniques. She puts her character on a psychologist's couch, reveals his childhood memories, and examines metaphors and context within the script. But there are many methods for understanding a character and the complex subtexts of a play. One of the many methods for rehearsing is a twelve-step model created by Ivana Chubbuck. It's popular among Hollywood actors, and Blokhus uses it too. The gist of the method is finding motivation in the piece as a whole, but also in each separate event. It's a memory-friendly method, because it works on the principles of memory: we try to understand events in our lives by tying them to our own goals and wishes. The idea is to flesh out the character by creating a backstory for them and making up inner monologues that would naturally have run through their head. Actors do this to get better acquainted with their character and, by doing so, enable themselves to express their character's feelings more genuinely. Neuropsychologists would classify this as a form of deep encoding, whereby memories are consolidated into a robust memory network. But, ultimately, colors and shapes are what hold it all together for Blokhus. Her synesthesia is a surprisingly useful tool for her as an actor.

"I remember the role because I go to an emotional place, and via my emotions, confront an existential problem where something in my own life is at stake. The colors and the shapes express the emotion to me and help me remember," Blokhus says.

For an outsider, it may seem incredible that a person can remember that much, a script that takes hours to perform. But is there anything even more difficult? What if you had to perform an hours-long script set to complicated music?

Johannes Weisser is an opera singer, and his job is to memorize three-hour operas, sometimes in languages other than his mother tongue. An opera singer who can't remember their lines will soon be out of work.

"I have no specific technique apart from approaching it from an artistic and musical perspective. I have to understand what I am singing; I have to know what each word means. I have to learn where all the pauses are in the music. But of course, the music and the director's instructions help me. As a rule, interpreting the music is more taxing than memorization. When that part is done, I have so many hooks to hang the words on that I know them by heart," he explains to us. He doesn't hide the fact that he sometimes has to cram in order to memorize the material. As he practices, he stands farther and farther from the music stand. "When I can no longer see the notes, I know it by heart."

He can sing *Don Giovanni* or *Così fan tutte* any time. He's nailed these two Mozart operas to the point that he only needs to flip through the music to find his role again. That's hours of music.

"The most difficult thing to learn is unchallenging music that offers no resistance, or music that doesn't interest me that much. Difficult things are easier: things I don't understand, or things that are challenging. The difficulties I encounter are the hooks that help me remember. For example, I'm working on the opera *Onegin* at the moment. I discovered pretty quickly that the piece poses totally new challenges for me, and that made me happy. Whenever I struggle with a part, I have to take the time to get it right, and that gives me something to remember it by."

There is more to this than simply cramming. Like Marie Blokhus, he approaches the work as a whole, tries to understand what the opera is about and how the music is assembled. Everything must mean something to him.

Weisser's and Blokhus's experiences fit with our knowledge of how memory works. Both the opera singer and the actor use their own memory to their advantage. When they understand whatever they sing or say, they can remember it better, because they've created a memory network. Connecting a role to one's own emotional life will strengthen the memory further. But what if you have to remember something outside of the way memory works best—that is, without a context? What if there's a lot at stake, and you have to remember many details without any history or emotions to tie them to?

"My understanding is that in the world of pub quizzes, people seldom or never use memory techniques. 'A gentleman quizzer never crams,' or so they say. Expert quizzers take pride in remembering things naturally. But people who are into quizzes read the paper with a notebook by their side and write things down. Every time I'm curious about something I read, I check with Wikipedia to remember it for later," says Ingrid Sande Larsen, three-time Norwegian champion in team quizzes who has also been president of Norway's quiz association.

The problem with cramming for a quiz is that you're supposed to know a little bit of everything. Where would you even begin?

"It's hard to remember things you're not interested in. Generally speaking, the quiz teams are made of people who are more curious than most, who are interested in all kinds of things," she says.

"I've discovered that many of the things I know best are from when I was ten and up through high school. I can hear a tune and know exactly where I was after school when I heard it the first time, who I had a crush on, and how things smelled and tasted. I know the hits from the 1980s and 1990s whether I want to or not," she adds.

In other words, she benefits from the reminiscence bump and her own personal autobiography.

She also remembers well things that have been asked in earlier quizzes.

"Right before the European Championships in Derby, England, in 2010, I had checked on Wikipedia to see who invented rugby. As it happened, I got the question and I nailed it: William Webb Ellis. I can still envision the room when I answered it: where I was sitting, and details such as the fake marble columns and the copper ceiling lamps."

As with the divers, the place where Larsen learned something became part of the context and helped her remember it in the future.

It seems that many people who rely on their memories don't use mnemonic techniques, nor do they cram. They're just passionate about what they're doing. Are memory techniques useful at all?

"Of course, if you have a good memory, you have no need for memory techniques," Oddbjørn By tells us.

By makes a living as a memory trainer. He has written several best-selling books on the subject and teaches courses for people who want to improve their memory. He is a Norwegian champion in the sport of memory and was, at his best, ranked twenty-second in the world. He participated in the World Memory Championships for ten years. Competitors in the championships take tests they can't possibly cram for in advance, such as memorizing endless lists of numbers they have to repeat in the right order, or remembering the order of a deck of cards they saw for only a few seconds.

"I envy cross-country skiers, who can work out with others in nature, while I sit alone indoors memorizing numbers off a sheet of paper," By says. "We have to memorize completely useless things, and it's difficult to find the motivation. That deck of cards, for instance; I am really tired of cards."

The best way to remember an entire deck of cards in the right order is to associate each of the fifty-two cards with a person or a particular object. Then when two cards appear side by side, they become part of a story. There are billions of possible combinations in a deck of cards, and the story that connects them can quickly become unlikely and bizarre. Every time By pulls the eight of spades out of the deck, his story includes the character he's linked to that card— Saddam Hussein. The seven of spades is a slave. If they appear side by side, it's a question of linking them together in a meaningful way, preferably as some kind of a story.

"You never know what story may emerge. Not long ago, for instance, Saddam Hussein had a baby!" By tells us.

His technique doesn't just work for memory championships. A more practical application is helping people with memory loss improve their memory. By calls it "artificial memory." He doesn't think he has an exceptionally good memory; he is just good at using memory techniques.

"People with memory problems tend to blame everything

they don't remember on things holding them back, like disease or old age. They forget that it's common to forget," he
argues. Forgetting—although we have healthy and quick
brains—is seen as proof of a flawed memory.

That's why many people won't use memory techniques;
they want to prove to themselves that they can remember
without them, that they're healthy and normal.

Some of By's techniques are surprisingly mundane,
such as writing things down, taking photos, or leaving his
umbrella in his jacket; there is, after all, no point using
all your energy to memorize things. But there are other
techniques made for memorizing things without using
external aids—phones, calendars, notebooks. These are
the techniques that memory artists use to wow audiences
in championships and make memorization appear almost
magical. The best known among these was developed
two thousand years ago by Roman orators and is called
the "method of loci," or the memory palace technique. It
involves placing items along a path within your mind.

Before we learn more about the method of loci, let's
head back to London for a bit. We've now given cab driver
Judy yet another task, but this one's so easy she could do
it in her sleep. From Short Street, we're driving to Shakespeare's Globe, the re-creation of Shakespeare's famous
theater that has been built by the Thames, and soon she's
opening the cab door for us on the left-hand side. But
though Judy pilots a machine driven by memory, she has
no idea that she has delivered us, in her black taxi, to the
biggest memory machine of the Renaissance: a theater.

We're now standing in the rebuilt theater with a group of tourists from Holland, Australia, the United States, and Denmark, the home of Hamlet. The theater was completed in 1997, four years after the death of Chicago philanthropist Sam Wanamaker, the man who financed the rebuild.

There's an open atrium in the center of the circular wooden theater, and in the 1500s when tickets were a penny each, the London rain spattered the cheeks of those standing in front of the stage. For two pennies, people could sit, safe and dry, up in the galleries. The roof is covered in moss, the sprinkler system clearly visible. The guide tells us that the roof, a fire hazard, was built with an exemption from the London fire regulations. There's a good reason why they're worried about fire. The original theater burned to the ground in 1613.

How did the Globe get its name? And why did someone bother painting the signs of the zodiac (Cancer, Pisces, Taurus, and so forth) on the ceiling above the stage, where only the actors could see them? The Heavens, they called it—a counterpart to what was taking place beneath on the stage, where even the devil could make an appearance.

Onstage this cold day, a group of children recite Shakespeare in high-pitched voices. It's a school class practicing the master's pieces. A visible white mist accompanies their words, evaporating in the gray air, before the children, laughing, disappear again.

They know little about how people saw the world in the Renaissance.

Picture yourself standing on the stage of Shakespeare's Globe Theatre—the original Globe—playing Hamlet. For a quick second, you glance up at the ceiling. Now you can remember where you are in the script, because above you there is a mnemonic aid—an astrological map painted yellow on a blue background.

Some do claim Shakespeare's theater was simply built as one big memory machine. The method of loci was developed long ago by the great orators, such as Cicero, to help them remember what they wanted to say from the podium. They attached specific images to the ideas they needed to remember and then placed the images along a familiar path they visualized in their mind—the road to the Senate, for instance. While they spoke, they could mentally walk the road and easily pick up the next point in their speech. The Renaissance thinker Giulio Camillo introduced what he called the "memory theater" based on this technique; his memory palace was a theater stage. The idea was refined by the Renaissance alchemist Robert Fludd, who considered the theater to have magical connections with our memories. He extended the idea to the zodiac, which he used as a memory aid; the way he saw it, humans were connected to the stars in the universe. Mastering the art of memory meant becoming a real magician with power over the stars. The modern Renaissance man stood in the center of the universe, under the stars, and via the magic of memory, he could move the world.

"Give me my Romeo: and, when he shall die," Juliet sighs from the stage of the Globe and glances at the star-sprinkled ceiling above:

"Take him and cut him out in little stars,

And he will make the face of heaven so fine,

That all the world will be in love with night,
And pay no worship to the garish sun."

The globe and the stars; the Globe and the ceiling above the stage. That was the world that would help the actors remember the piece they were performing and, at the same time, connect them to the universe and all the magical powers it contained.

Today, people don't consider the method of loci to be magical, but it's still a very useful mnemonic technique. It relies on two important factors. One is picking a familiar setting for your memory palace, and the second is choosing significant images to represent the things you want to remember. Using a well-known place saves space in your memory and provides a natural order to what is to be remembered. Imagine, for example, that your path takes you from your home to your school (even adults remember well their way to school, though many years have passed since they last walked it). You choose some typical stops along the way: a bus stop, the yellow house down the street, an intersection, a corner store, and so on. Then you pick keywords to signify what you have to remember at each stop—but the keyword must be reworked into an evocative picture, creating a unique memory trace. A good example of something one might memorize this way is the periodic table. Imagine hydrogen (a big part of water) at the first stop: the bus stop has flooded! People cling to the bus stop sign, they sit inside their umbrellas, using them as boats (not the least bit probable, but at least distinct!). Don't fall for the temptation to use the little puddle that was there before, thinking it will help you think of hydrogen. Things that might be related to what we need to remember but were already there, part of the environment, just fade into the background and disappear. The next stop is the yellow house down the street. There you find helium, in the form of an enormous bouquet of helium balloons attached to the house, almost lifting it from the ground. The next stop is the intersection, and there you will find lithium in the form of a giant container of lithium batteries in the middle of the intersection, so big the cars have to detour around it. The memory palace technique can be used to memorize ordinary things too, like shopping lists, the keywords of your history syllabus, or chore lists.

"I have about one hundred different memory palaces because if I am supposed to remember different things, I have to vary the route so I don't mix things up. My favorite palace is my brother's barn. I know it well, and it has many nooks and crannies," Oddbjørn By says.

As a student, he used his memory to pull a stunt that aggravated his fellow students. Without enrolling in a class, he registered for the final exam and memorized the coursework in only two days. His secret was the method of loci.

"A good friend of mine was studying extinct religions, and I borrowed his well-structured notes. I placed the material in mind palaces and was able to remember it for an oral exam," he recalls. He got a B in the subject, and that's what irritated the others. Why would anyone struggle for months at a time learning a subject if it was that simple? But today, after twelve years have passed, By doesn't remember much about the Mesopotamian gods. But a history course, on the other hand, which he diligently studied for an entire year, has stuck with him, though his grade was not too impressive.

"I don't recommend that people study for a course in this superficial manner, but the memory palace technique can be combined with other, deeper study techniques," he says. "If you happen to have some extra time, and nothing to fill it with, why not use it to learn something new?"

Creating a memory palace may sound difficult at the outset, but it can be used in many different, even stressful situations: for the exam, on the stage, and even on the dance floor.

"The technique works well if you're doing something meaningful while, at the same time, walking the route in your mind; it only takes a couple of milliseconds. For actors and opera stars, this means picking out keywords from a script and organizing the rest of the text around them. Then you place the keywords along your route. The more of an expert you become, the fewer keywords you need to remember the material; understanding the material helps you remember it better. Many memory techniques work better if you already know some of what you are about to learn."

Even experts sometimes use memory techniques. Solomon Shereshevsky, the man who couldn't forget, gradually turned to the method of loci. In the beginning, he used his own self-made method, whereby he simply visualized the things he had to remember positioned along the streets of Moscow. He didn't need to create particularly distinct pictures because he remembered well without them. But sometimes he missed something because he had placed a memory in the shadow between two lampposts. He simply walked on by without noticing the unassuming little egg he was supposed to remember; it had faded into the background because his imagination was almost too vivid. When this happened, he realized that he had to structure his memories better, along a single route, using symbols.

In addition to the method of loci, you can use memory rules: for example, first-letter rules, creating a word from the first letter of each thing you want to remember. You can use the orange trick: put an orange in your bed when you need to remember something important. When you return home, you'll see the orange and remember why it's there. There are, of course, mind maps. Then there are flashcards—index cards with information on each side; countries might be on one side, capital cities on the other. Go through the pile. Those you know, you put to the side; others you drill through again until you know them.

In order to benefit from these methods, you often need to understand the topic at hand. You can't make mind maps without having familiarized yourself with the material and mastered it. The methods are basically all about making learning easier. They provide motivation. Cramming becomes unnecessary when you can learn in a more memory-friendly way. But you can't get around the fact that learning, as a rule, takes some effort. Take, for example, the periodic table, anatomical names in medicine, Latin names in botany, mathematical formulas, grammar rules in languages. Before they become part of you, you have to hammer them into place—or walk them through a memory palace.

"Mastering memory feels good, even more so for people with memory problems," By concludes.

Before we leave, he is going to demonstrate one of his tricks: memorizing a list of numbers. At his best, he was ranked ninth in the world at this. He asks us to read a random list of thirty-five numbers, one number per second, paced as evenly as possible. For us, no memory champions, the natural thing to do would be to repeat in our heads as many numbers as possible, right away, before they disappear. It's likely we'd remember only a few from the beginning and end of the list. People with average memories can remember six or seven numbers; that's how much our short-term memories can store. When we finish reading the numbers to Oddbjørn By, though, there is silence.

The Norwegian memory champion leans forward, his hands over his mouth and nose and a distant look in his eyes. Has he forgotten all the numbers already? A minute passes; two minutes. Then he begins to list them off. The first numbers are correct; he asks to skip a few numbers that he saves for last, and goes on to recall more numbers from memory. He is like a magician, pulling one rabbit after the other out of his top hat. He is so quick that we have trouble keeping up with him. In the end, he manages thirty-four of thirty-five numbers correctly. It is almost magical; it shouldn't be possible. But his method is to arrange the numbers, in pairs, into the shapes of creatures, which he then places in a memory palace. Two by two, the numbers become zebras and goblins, people and animals along his route. In principle, this is a method anybody can use, but if you want to be the best in the world, you have to practice a lot and stay focused. Perhaps you also need a certain amount of talent, as in sports. Those of us without talent can remain happy memory amateurs.

"Working with memory has changed me. When I read a book, there are more images in my mind. I have become much more visual," By says.

Anders Fjell, whom we met earlier, and Kristine Walhovd together lead the Centre for Lifespan Changes in Brain and Cognition at the University of Oslo, where they are involved in a number of research projects. Their goal is to reveal what influences memory throughout life.

One of their projects is a comprehensive study of the effects of memory training. Their subjects don't include any exceptional taxi drivers. Instead, they're training the memories of two hundred senior citizens. Do memory techniques make a difference in a group of seventy-year-olds?

"After ten weeks of memory training, the group of seventy-year-olds remembered as well as people in their twenties who had not learned any memory techniques," Walhovd tells us. Because the older participants are so determined, they usually make a solid effort. They get more out of the training than one might expect.

"They know they must make an effort and take the task seriously, perhaps more so than the younger ones," she asserts.

The changes in their memory skills are also visible in their brains. Just like Eleanor Maguire, Walhovd and her colleagues scanned the brains of the volunteer memory trainees and saw physical differences. But even if our brains change, improving our overall memory is still elusive, because it doesn't look as if training helps us to remember more broadly.

"If what you practice is memorizing one hundred words in the right order, then that's what you become good at: memorizing words in the right order," Kristine Walhovd says.

But what about people whose memories are poor due to brain injury? Completely regaining your memory after a serious brain injury is too much to hope for, especially if the hippocampus is affected, or if there are large and long-lasting injuries to the brain. The goal of rehabilitation is typically managing everyday activities better. Sometimes this means using aids (like a day planner, diary, or calendar with an alarm function), implementing steady routines, writing shopping lists, and recording messages. Changing the way one remembers things can be a long process for someone with a brain injury; sometimes it is also upsetting. Having memory problems means spending a lot of extra time and energy mastering tasks and learning new skills. Sometimes rehabilitation is as much a process of figuring out one's own limits as it is active memory training. For someone with a brain injury, memory techniques can help regain a sense of control.

According to research—that of both Eleanor Maguire and the research duo Walhovd and Fjell—memory training is not the same as improving memory in general. But remembering better with the help of certain techniques is nothing to sniff at if you can make it work.

Mastering memory techniques makes the brain better equipped—at using memory techniques. This is true for the taxi drivers, whose spatial memory improved while their hippocampus got bigger, as well as the grand masters of chess, who are better than everyone else at remembering chess positions, but not much else.

CHESS GRAND MASTER Jon Ludvig Hammer leans over the chess board. Then he bounces up.

"Are you kidding me? Are you frigging kidding me?" he says and laughs, astonished. "The next time, you have to give me a warning!"

It's just as if we held a glass of sour milk under his nose—the reaction is that intense. We have shown him one of the nonsense boards. He instantly saw what it was, and now he fumbles to place two correct pieces on his board, shocked and overwhelmed by the chaos we have shown him. But when he gets to the next nonsense board, he has developed a strategy and focuses on a limited number of pieces; he ultimately manages to get nine points on the last one.

On the proper boards, he makes a maximum of four or five mistakes. The Carlsen–Anand board is immaculate; he gets all twenty-three pieces correct after studying the board for five seconds.

In the amount of time it takes us to say "Viswanathan Anand, Hikaru Nakamura, and Garry Kasparov," the time it takes to glimpse a seahorse in the water or decide to turn left at a confusing intersection in London, the time it takes to walk through Oddbjørn By's mental barn, Jon Ludvig has managed to identify all the pieces on the board, where they belong, and also what game of chess this board is from. It is superb.

But for some of the boards, some pieces do end up in the wrong places. He did, after all, have only five seconds.

Jon Ludvig refuses to leave. He remains sitting there. He is obviously tortured by the fact that he wasn't completely right the first time and insists on one more chance.

"I'll set out the four real boards correctly in the right order, and this time I will get all of them right. Without looking at the boards again, of course," he says.

We realize we can't stop him.

He is very fast now. When board number one is completed, he doesn't bother to stop or take the pieces off the board—he just continues, as if in a trance, like a spiritual medium of chess, channeling kings and queens. He sprinkles all the pieces into place. On the four boards, he places ninety-six pieces. All are correct, except one, a pawn.

"The pawns make up the skeleton. I build everything around them. I structure the board logically, starting with the pawns," he explains.

His strategy for remembering the board does not remind us of Olga's, for whom the middle row was the most important.

Jon Ludvig Hammer is a full-time professional chess player. He can spend ten to twelve hours a day practicing opening moves in chess; that's his job. He has read all the books Simen Agdestein showed us. He has crammed attack moves and response moves so they are all available to him when he is in a tournament, ready to deliver. Further into the tournament, memory may play tricks on him as he becomes tired.

"Sometimes I forget what strategy I was using, because I have been thinking so hard about an alternative strategy in the meantime," he says. And, because working memory doesn't have the capacity to hold that many things, just going to the bathroom can make him lose his whole train of thought.

Jon Ludvig stares at his last board. His hand darts over a white pawn. He picks it up, puts it down again.

"There is a piece missing," he says. "But it wasn't there when you showed me the board the first time. There should be a pawn here, to defend the knight. This whole area is now open in a weird way."

We do a double take and check again. We had simply forgotten to place a white pawn on c2. The game has most definitely turned. It is checkmate.

THE ELEPHANT'S GRAVEYARD

Or: The art of forgetting

I stand amid the roar
Of a surf-tormented shore,
And I hold within my hand
Grains of the golden sand—
How few! yet how they creep
Through my fingers to the deep,
While I weep—while I weep!
O God! can I not grasp
Them with a tighter clasp?
O God! can I not save
One from the pitiless wave?
Is all that we see or seem
But a dream within a dream?

EDGAR ALLAN POE,
"A Dream Within a Dream"

B ERLIN, 1879. THE city's most prominent citizens promenade beside the river Spree. Along Unter den Linden, they sit at outdoor cafés, enjoying the warm weather and the blossoming linden trees. They rearrange their dresses and their top hats and breathe in the spring smells: horse manure in the street and fresh-baked pretzels. The abundant foliage casts shadows on the ground.

"What a wonderful time!" these bourgeois may be thinking beneath the tree canopy in Berlin. "I wonder if this particular moment will stay with me for life. Will I remember the breeze making the linden trees sway, when I think back on this a year from now, five years, twenty years? How much of this will I forget?"

Meanwhile, in a laboratory at the University of Berlin, a lone researcher is about to begin a groundbreaking experiment. He is going to attempt something never before tried in history. He is not going to conquer a mountain, invent the lightbulb, or travel to the Moon. Nobody in high school history class will read about what he is going to do. But in the history of psychology, he will be hailed as a great hero, a man who walked where no one had walked before. Hermann Ebbinghaus will be remembered forever for his efforts to do something completely ordinary—forget. While Berlin's upper classes stroll the river in the spring sun, Ebbinghaus feeds his memory with meaningless syllables. BOS—DIT—YEK—DAT. He studies them intensely and tests himself, hour after hour, day after day, until he can repeat every list of twenty-five nonsense words in the right order. While life unfolds outside the University of Berlin,

Ebbinghaus immerses himself in syllables. He's chosen to study empty combinations of letters because they are completely free of the troublesome contamination of emotions, ideas, and his own life. He tests how much he can still remember after a third of an hour, an hour, nine hours, a day, two days, six days, and thirty-one days.

He wants to find out how fast he forgets; it's as simple as that. Sure, outside of psychology circles, this may not seem like an accomplishment worthy of much celebration. We can plant a flag on the South Pole, but we can't do the same with the act of forgetting—we can't discover it and declare, *Here it is!* While Solomon Shereshevsky could make a living and earn applause by remembering superhumanly long lists of words and numbers, nobody would pay a nickel to see Ebbinghaus stand onstage and *forget.* It's safe to say that he had undertaken an unglamorous task. Though what he did wasn't particularly exciting on the surface, it was actually quite sensational. Psychology was a brand-new field of research; nobody had researched memory like this before. Up to that point, measuring thoughts wasn't something anyone could have imagined. But Ebbinghaus performed such a significant feat that scientific society was forced to take him seriously.

It was a demanding task to document forgetting. Ebbinghaus didn't want to leave anything to chance, so he did all the experimenting on himself—and, really, who else would have agreed to the job? That way, he could trust that he was in full control of all variables. It meant he also had to keep his own personal life in check, so no sensational memories could influence the impersonal, scientific building blocks he was memorizing. After several years of intense and one could say *ascetic* work memorizing and forgetting, Ebbinghaus published the book *Über das Gedächtnis* (About memory). Up until 1885, memory had belonged to the realm of philosophers, writers, and alchemists. Never before had science focused on forgetting. How, then, do we measure a vanishing memory?

If Ebbinghaus memorized a list of nonsense words and after a while—let's say after one day—could come up with only a little more than half of them, had the rest been forgotten? Yes, there and then, some words were forgotten, and the difference could be measured and called forgetting. But this was not thorough enough for Ebbinghaus. It could have been that the words were still stored in the brain and only the access to them had been weakened, so that he couldn't reach them by will. It could be that, deep down, there were remains of memory traces that could be wrung out like water from a wet cloth.

"We cannot, of course, directly observe their present existence, but it is revealed by the effects which come to our knowledge with a certainty like that with which we infer the existence of the stars below the horizon," he determined.

He chose to approach forgetting from another angle. If he had forgotten a list of meaningless words, how long would it take for him to relearn them after some time had passed? For every new learning effort, he measured how many repetitions, or seconds, were required for him to remember the list again. If the list had been completely forgotten, so that not a single strengthened synapse remained, relearning the list would take him as long as it did the first time around. But if he'd retained anything, it wouldn't take as long to relearn it. In this way, he calculated the natural course of forgetting and discovered that our memories disappear most quickly in the first hour. After a day, more has been lost, but the process of forgetting quickly slows down, so that after a month, we've forgotten only slightly more than after a week. His research led to what we today call the *forgetting curve*. Shown as a graph, it descends quickly in the beginning and then tapers off.

Never has any researcher exposed his own weakness—his forgetfulness—for the benefit of humanity with such intense dedication as Ebbinghaus. For several years, he wrote page after page about what he had forgotten and tracked forgetting with tables and numbers, satisfied to have contributed to the science of psychology. Maybe he would have preferred to be in the streets of Berlin enjoying the spring sun, sipping a cup of coffee with friends, and strolling slowly along the river. But he wrote nothing about his personal memories from the time of the experiments—except that he strove to keep personally meaningful experiences to a minimum, in the service of science.

What Ebbinghaus proved was that memories, when they don't have anything to do with ourselves or what we care about, gradually wither. But he had no way of understanding exactly *what* crumbles in our brain. The creation of memory traces was, as we've said before, not proven until the 1960s by Terje Lømo. Memory traces probably weaken over time. It seems as though, unless we practice and maintain knowledge until it's become firm in memory, the neurons involved in remembering eventually return to their original state. This is probably a good thing. It gives the brain space for new memories. The other thing Ebbinghaus revealed was that the brain begins tidy-up work shortly after new experiences have entered memory. This too is probably a practical trait in memory. It's better to clean up sooner rather than later. And it ought to be obvious fairly early on whether or not an experience is important enough to be stored. When Ebbinghaus researched forgetting by measuring learning, he also made it clear forgetting and remembering go hand in hand. They are two sides of the same coin.

If we don't forget, the storage space in our brain fills up (Solomon Shereshevsky and his like notwithstanding). For most of us, some memories have to depart to make room for new, perhaps more important ones.

"If we remembered everything, we should on most occasions be as ill off as if we remembered nothing. It would take as long for us to recall a space of time as it took the original time to elapse, and we should never get ahead with our thinking," William James pointed out in 1890.

Still, forgetting is something we fear. Forgetting is aging; it's decay and impermanence, a memento mori. When the days pass and we cannot remember them, it means we're one step closer to the end of life, without anything to show for it.

That's why blogger and author Ida Jackson has kept a diary since she was twelve. "It feels like it's helping. I lose fewer things that way. If I look up a certain day in my diary and see that we had dinner with friends, then I remember more of that dinner, even if I didn't write down any details." Ida is a collector of memories, scared to death that the moments will be gone forever.

Generally speaking, forgetfulness reminds us that we are not in full control. It *is*, after all, impractical to forget appointments, friends' birthdays, phone numbers, and everyday experiences. Forgetting names can be embarrassing. But forgetting is much more common than the most intense hypochondriacs like to believe, and it is seldom a sign of dementia or early Alzheimer's. Lack of sleep and general exhaustion are enough to cause important things to slip.

Even when our brains work perfectly, most of us forget more than we'd like to. We forget names, because they usually don't have a logical connection to the person they belong to. Among the Vikings, a thousand years ago, it was not uncommon to get named after one's physical characteristic or personality. Some common Norwegian names still in use today originally meant "cockeyed" or "spinning"— fitting of someone with more energy than the average person. Some family names, like Smith, originally described a profession. Today, though, names are often random labels without a clue to the physical person they designate. Only through repetition and association are the names attached to the person in our memory.

We forget faces, because they are complex and difficult to describe. The small part of the cortex that specializes in perceiving and remembering facial features helps us navigate our social worlds by quickly processing the faces that we meet. But much like other brain functions, this doesn't work perfectly. When we first recognize a face, we don't necessarily remember who it belongs to. We forget where we know someone from because when we first see them, we haven't activated the memory network we originally placed them in.

Faces and names and appointments and telephone numbers, your sister's birthday or a bill that's past due: Where does all this everyday forgetfulness come from? Forgetting isn't just the process by which memory traces fade away. Forgetting takes place at all stages of memory: encoding, storing, and retrieval. Often experiences never make it into memory at all. In order to get to the stage where memories are stored and consolidated, experiences first have to go through a screening process.

The first obstacle is attention. Magicians and pickpockets thrive on exploiting attention, because they know it can focus on only one thing at a time. While you're looking at the map the thief pushes into your face to ask for directions; his hand sneaks, unnoticed, into your bag.

In 1970, a Norwegian TV reporter stopped people in the street and asked them some pointless questions in front of the camera. In the middle of the interview, someone walked between the reporter and the interviewee, carrying a large plywood sign. While the reporter was out of sight, a comedian, Trond Kirkvaag, took his place, wearing either fangs or a king's crown. The people being interviewed didn't seem to notice that anything had changed. One of them even pointed out a mistake in the interviewer's question, even though the person he told was someone entirely different from the person who asked the question in the first place. This was, of course, a comedy show on television, but it revealed a truth. If you are interviewed on TV, your attention is on the microphone in front of you. With all the adrenaline flowing through your body, you might not even notice if the person interviewing you is suddenly replaced with someone else.

Approximately twenty years later, Harvard researcher Daniel Simons did a similar experiment, which later made him famous in psychology circles and won him an Ig Nobel Prize. He made a movie—perhaps the world's most boring movie, if you're used to Hollywood fare. It shows six people passing a basketball. Those watching the film were asked to count how many times the ball was passed by the people wearing white. Half of those who watched the movie proudly reported that the ball was passed fifteen times, but when they were asked if they had seen a gorilla, they were insistent that they had not. Still, if you watch the film, you'll see a man in a gorilla suit slowly strolling among the basketball players, stopping and beating his chest, before showily spinning on his heel and exiting to the left. Our attention is like a camera lens; everything outside its focus is blurry and ends up in the background. We can hardly call this sort of thing forgetting; the only way the experience affected our mind was via a brief burst of sensory stimulation that went unnoticed by the rest of the brain.

The next obstacle standing in the way of something becoming a lasting memory is our working memory, aka short-term memory. It may be memory's weakest link, and its most critical one too. There is only limited room here, and memories can stay for only a short time, about twenty seconds. Henry Molaison still had a working memory and could maintain a conversation as long as he had a meaningful connection to the topic. As soon as his thoughts wandered, the conversation was over. This was the healthy part of Henry's memory. But his memories never got any further, into long-term memory. More often than not, our experiences end up like Henry's short-term memory, eluding further storage.

When psychology professor Alan Baddeley performed his now-famous diving experiment off the coast of Scotland, he had already started on another research project. It was actually this *other* project that made him a giant in psychology and has made waves ever since. It involved trying to understand what happens to the fragile, volatile memories of the here and now.

In the 1960s, Alan Baddeley had worked for the General Post Office in Britain, creating a memory-friendly system for postal codes. Unfortunately, the system was never adopted, but knowing how memory allows us to retain random numbers for the short time it takes to write them on an envelope raised many questions for Baddeley. It inspired him to research the topic of short-term memory further, together with his colleague Graham Hitch.

You'd think short-term memory would be easy to understand. Either we remember something for a short time, or we remember it for a long time. In the 1950s, researchers figured out that our short-term memory has room for seven units of information at a time. They called it the "magic number seven" (later revised to the not-quite-so-catchy "magic number seven, plus or minus two," taking into consideration normal variation in individuals). But Baddeley and his colleague soon discovered that short-term memory is much more complex; it is an active process, *working* memory, not a magical container. They also found that working memory contains several stores, each with its own specialty: linguistic information, images, episodes from life—perhaps even more layers, each tied to a sensory modality, such as smell, taste, and touch.

"I remember one of our earliest experiments, one that led us to draw up our model of working memory, the topic that has occupied my mind for the past forty years," Alan Baddeley shares with us.

"We asked volunteers to remember sequences of five words that were similar in sound, such as man, cat, mat, can, and hat, and then we compared their performance with that of having to remember dissimilar-sounding words; for example, pit, day, hen, pot, bun. There was a clear difference: the dissimilar-sounding words produced up to 90 percent correct responses, as compared to only 10 percent of the similar-sounding word sequences!"

They had discovered that working memory has a separate store for spoken language, called the *phonological loop* (try remembering *that* till the end of the chapter!), whose only task is to store linguistic units.

"This is the part of working memory that allows us to learn a foreign language, for instance," he says.

Our ear picks up on new, not yet understood words, which our cortex interprets as language sounds and sends to the phonological loop, where they are held automatically for a few seconds. From there, these sounds can be repeated, in a loop (a process that is automatic but can also be voluntary). If we manage to repeat them for a long enough time, they may stick in our memory and we can say we have learned something new. Things we hear from teachers, spouses, customers on the telephone, or TV advertisements enter the phonological loop and compete for space. They say that messages can go in one ear and out the other, and this is an apt description of working memory. It is the place where our stream of consciousness is trapped and held, for a brief moment, in front of our inner eye and ear.

Visual information is processed by another part of our working memory, and the two systems can function fairly independently of each other.

"Regretfully, there has been less research on the visual part of working memory," Baddeley admits, "although it's currently a very active area."

Several of his experiments have shown that when subjects simultaneously absorb visual stimuli and words, their ability to remember words is less compromised than when they see several words at the same time. In other words, we can juggle several types of information without our memories suffering. Everything is handled by a *central executive*, which steers the attention to where it is needed, prevents awareness from drifting, and keeps undesired information out of working memory.

During the forty years Alan Baddeley has been researching working memory, some new discoveries have changed the model for how here-and-now memories work. One of the latest additions to the model is the *episodic buffer*. It acts as an intermediary between our attention and our memories and thoughts, fetching them from our long-term memory and presenting them to us in the here and now.

"Think of it as a TV monitor, where thoughts, memories, and images are being shown to us," Alan Baddeley explains. "It's a passive monitor that plays a multidimensional show, a show that has been prepared for us elsewhere in the brain and then projected on the monitor."

Behind the curtain, the brain busily works to prepare the show for the screen. Working memory is where we think, solve problems, do math. It is also where memories are acted out before our inner eye.

The working memory model is useful for understanding how and why certain things never enter our memory at all. Forgetting in working memory is something completely different from forgetting in long-term memory.

Working memory is set up to hold information for a very short time; it provides only temporary storage. It's like a mail shelf, where employees are supposed to pick up today's mail so there's room for new deliveries. The only difference is that on this mail shelf, if you don't pick up your mail on time, someone throws it out. It is normal to forget in this way. It's a natural part of having a human brain.

"Forgetting is an important aspect of memory—it helps us see what's important," Baddeley reminds us.

Forgetting is so central to remembering that we almost take it for granted. Still, many complain about their poor memory, even though their ability to store new memories is totally normal. They are just victims of their own natural working-memory screening. The situation is worse for people with attention deficit hyperactivity disorder (ADHD), whose problems with attention make it harder for them to focus on things long enough to store them.

We often forget things when other thoughts demand space in our working memory. Worries are a classic example. Things we worry about are things that are important to us; they overflow with emotion, crying for attention. That's why they are sent directly into working memory.

Here's an example: you are studying for an exam; you're afraid of flunking. You are trying to focus on maritime ecosystems, but it's a hard struggle because you're so worried, and your worries are clogging up your working memory. The life cycle of plankton competes with thoughts like "If I don't pass the exam, I'll have to do the class over again, I'll lose half a year of studies, I won't get to go to Greece this summer, I'll have to find a summer job, I'll be broke, I'll never get a job, my parents will worry and nag, my friends will think I'm a loser, and they will go to Greece without me!" How many millions of plankton have to step aside to make room for these worries? The plankton lose. Even if you start off with a passion for plankton and their significance to ecosystems and the climate crisis, they're now flushed out to sea, far beyond your reach.

Working memory may also be to blame when we don't remember people's names the first time (maybe also the second and third). As we're shaking hands, the name we are supposed to remember competes with all the other thoughts going through our minds: how we look, what we'll talk about after the handshake, whether we grabbed their hand too hard or too soft. Many of us are afraid we'll appear impolite if we don't immediately remember a name. It may, however, actually be a sign of interest in the person. It is above all the individual and what they stand for—not their name—that occupies our working memory during that first handshake and the minutes that follow.

Even those with extremely good memories may sometimes succumb to the failures of working memory. Compared to most people, Norway's memory champion Oddbjørn By has a completely different standard when it comes to forgetting. What if he forgets just *one* of the numbers he is supposed to remember in the exact right order? It would be disastrous! During the World Memory Championships in 2009, he was at the top of his game. Beside him, though, sat another contestant, a Chinese memory master with a throat infection. He was coughing loudly. Noise is something By trains to handle by sitting in noisy cafés. On the day of the competition, though, his nerves got the better of him. A cough from the competitor after digit number thirty-seven sealed his fate. Even though he missed only that one number, the thirty-eighth in the row, he wound up with thirty-seven points out of a possible one hundred, his life's most disappointing placement.

Yet another form of forgetfulness becomes evident when we're retrieving something from memory. To remember, we normally rely on cues to take us to the exact memory. They make it possible for us to find the right memory network, the particular fishnet of memories, and haul in the catch. Sometimes the cues get mixed up. We latch on to the network of something else that looks similar and steals our cue. It is a bit like googling: we must use the correct search term to get a relevant hit in our memory. And when the results of the search appear, we have to choose one among many.

Mnemonist Solomon Shereshevsky could remember meaningless lists of numbers and words for an almost immeasurable length of time. Yet he still had a memory problem: he was afraid that all the things he couldn't forget would disrupt his ability to remember other things during his performances. In other words, he was afraid of remembering the wrong list of words! Even if after every performance he wiped clear the board where an audience member would write the list of words, they were still almost permanently etched into Solomon's mind. He made repeated tries to forget the list, but the more he tried, the more the words stuck. His solution was to imagine the words on a piece of paper and, in his mind, crumple it up and throw it in the garbage. We're uncertain if this actually helped him forget, but at least it tagged the list, separating it from the new lists he was trying to repeat, onstage, with everyone's eyes on him. Ironically, he had used his amazing memory to help him forget.

Forgetting names and messages in droves is one thing. All of life's little experiences running like sand between our fingers is something else. Remembering the important things in life is what really matters, isn't it? What's the point of spending a bunch of money on a vacation if we can't remember any of it afterward? Forgetfulness is a friend, sifting through everything to reveal the high points, the pearls in our necklace of memories. Most of what we experience does disappear. All those times we've waited for the bus, the trips to the store, the afternoons on the sofa; they aren't supposed to remain in our memory. Forgetting even touches memory's shiniest pearls. It leaves behind only an outline; the rest is reconstruction. It's how our memory remains flexible.

The most widespread kind of forgetting, the kind that affects our personal memories to the highest degree, is the one we all experience after childhood. Researchers call it *childhood amnesia* (or *infantile amnesia*).

Most of us have a boundary, somewhere between the ages of three and five, that marks the beginning of life as we remember it. Some remember further back, from around the age of two; others may have very few memories until the age of seven. Up to that point, it's a complete blank.

We remember our first years only through stories our relatives tell us. Why do we forget our early childhood like this? How can that boundary appear at a certain age? It's a mystery, one researchers have struggled with for well over a century—and a riddle humans have probably pondered as long as we've philosophized about our own awareness. It is, after all, a universal, obvious gap in memory.

There have been many theories. Does it have anything to do with the development of language? In the 1980s and '90s, many suggested that childhood amnesia could be blamed on children's lack of language to express their experiences to their parents and to themselves, making them unable to attach their memories to words. This assumes that language, when developed enough, is what makes it possible for us to remember things. But children who have just learned to string words together in sentences are able to tell us about things that happened earlier in their lives, before they had language skills, so this cannot be true. Perhaps when our language skills reach a certain maturity level, this reorganizes our memories? Is everything shuffled and moved into new linguistic shelves and drawers? Memories are stories, and proper stories provide structure to memories, right? But this doesn't ring true either, because then there would be a sharp difference between the character of memories before and after the linguistic reorganization.

Up until the 2000s, the question wasn't being asked the right way. From another angle, you can ask: When do our first childhood memories *disappear*, becoming part of our childhood amnesia? Earlier, people speculated about how, as adults, we remember childhood. However, that's not where we'll find the solution to the mystery. Our memories go through too much as we grow up for an adult perspective to be useful in understanding small children's memory.

At Emory University in Atlanta, psychology professor Patricia Bauer has set up a kind of children's memory lab, which proudly bears the nickname "Memory at Emory." Bauer wants to follow the natural course of children's memories, which demands patience and effort. To achieve her goal, she has to standardize the children's memories so the memories can be compared across age. When children come to the laboratory, they're given a certain set of toys that they don't have at home and are shown how they work. When they return to the laboratory a couple of months later, the children usually start playing with the same toy if they remember their last visit. This way, the researchers don't need the children to tell what they remember; they show it. As they grow older, their memory is measured by what they tell the researchers.

Patricia Bauer has completed many studies following the children's memories over time. Starting from the moment a memory is made, she can watch it take shape from the other side of the childhood amnesia boundary. The memories don't suddenly disappear when the child turns four. When you think about it, this is obvious. We know that three-year-olds can tell us about their summer vacation half a year later. Even two-year-olds are able, using their limited vocabularies, to talk about things that happened several months ago. Childhood amnesia doesn't just suddenly appear in a four-year-old who can still recall what happened a year ago. What Bauer discovered was that memories that later disappear into childhood amnesia are still accessible for children several years after the age of four, before they gradually fade away. To understand the process, we need to study the life span of a memory created in a two-year-old, a three-year-old, a four-year-old, and so on. A two-year-old's experiences become memories that last for a shorter period of time than those of a three-year-old. It seems as if our earliest memories come with a best-before date. They are perishables and degrade quickly. As a child grows older, their memories come with more generous best-before dates. Finally, as they mature, their memories reach the almost unlimited durability of canned goods. In hindsight, childhood amnesia sets in around the same point that our memories achieve adult durability. The memories made before that age become weaker and weaker until, at about nine years of age, they completely disappear in most children. The way language reorganizes memory doesn't explain why memories vividly described by a six-year-old vanish by the time that same child has turned nine. But language does have some effect. Bauer has seen a clear connection between how parents talk to their children about their experiences and how well those memories stick. Anecdotes parents bring up over and over again become part of a child's life story and come to life with the help of constructive memory.

"Everything you want your children to remember, you must talk to them about," brain researcher Kristine

Walhovd says. "As parents, we of course emphasize the positive experiences of our children."

This is how parents can contribute to their children remembering a good childhood.

"They say 'it's never too late to have a happy childhood.' A lot depends on how you weigh the episodes in your child's life," Walhovd adds.

There is great variation in how far back our earliest childhood memories go. Some are undoubtedly glimpses of real experiences. Bright flashes of light, sound, and, most often, a certain mood. Some claim that they have vivid childhood memories from before the age of two. Many have childhood memories that can be traced back to photographs they've seen or stories they've heard from family members. Our memory's reconstructive process takes these stories and images and brings them to life, even when there is no trace left of the original memory. In that way, a "false" memory is created of a real-life experience. These constructions may appear early on in life, stay with us, and feel like "real" memories. Over the years, we may easily forget that we were once told the anecdote. The act of listening isn't as memorable as the story we hear. If your mother told you about a family vacation you went on at, say, the age of two—and you are now five—it may well happen that you pictured it vividly and remembered the reconstruction, without remembering that it came from your mother. Your mother may well have forgotten herself that she told you the story. In this way, reconstructions sneak in seamlessly among our childhood memories. People often insist that their childhood memories couldn't possibly originate from other people's descriptions or photos. And it's impossible for us to tell.

Let's dive back through the cortex to visit the seahorse in its temporal lobe. Could it hold the key to the riddle of childhood amnesia? One theory is that the hippocampus isn't mature enough in early childhood to consolidate memories for good. Because not only does the hippocampus have to grow and develop, it also has to build a network with the cortex. This happens at the same time the cortex is going through a growth spurt. All this chaos probably results in memories that are less securely stored compared to how they will be when everything in the brain is properly in place.

Some newer, more sensational theories look at other aspects of the hippocampus's development, theories that are more speculative. Some claim that place and grid cells, which we know are central to our spatial memory, aren't ready to map the environment until a child starts to move around independently, and even after that, that it still takes some time before the system is done "calibrating." Still others speculate that the solution lies in a microscopic layer of protein called the *perineuronal net* that gradually envelops neurons and synapses. This fine-meshed protein net may protect the links between neurons, allowing the memories to more easily adhere. More and more netting is added to the brain throughout life. The downside may be that people's mindset hardens, so to speak, as the perineuronal net solidifies around the neurons. Our first memories can't get as firm a grip, because the perineuronal net hasn't been developed yet. However, this research has, so far, looked only at the brains of mice and rats.

Later in childhood and early adulthood, most of us are blessed with an observant personal memory that, as we've said, preferentially collects experiences in the reminiscence bump: many of the significant, new, exciting, sad, and transformative experiences that become part of our personal autobiography. Impressive as memory may seem, there's no denying that it goes hand in hand with obliviousness. Sometimes, it feels as if the days slide into a black hole as time passes. Is it possible to stop this? Can't we prevent forgetting from devouring all those little experiences? Think back on the last six months: perhaps, at first, there is only a vague outline, punctuated by holidays, birthdays, and trips. Then there is a virtual parade of the most prominent unique events. Everything is shortened, compressed. The air has been squeezed out, like clothes you roll tightly into a carry-on to avoid extra fees.

What happens if you fight natural forgetting? If you put effort into remembering all the unique moments that make up your life? Can the parade get longer? Can we put a lid on the black hole of forgetfulness, using memory techniques to remember the important events?

The thought is so absurd that it *has* to be tested. So one of us sets about trying to fight forgetting. For several months, we have interviewed memory researchers and actors and chess players, but the two of us each have our own approach for remembering these events. One of us submits to the organic nature of forgetting, while the other one tries to nail each day to memory in order, for one hundred days.

Let us, the authors, invite you to our memory theater for a short performance—we'll set the stage for a talk show!

Hilde will play the role of the talk show host. Ylva is the main guest who will entertain viewers with the tale of how she pretended to be Ebbinghaus for one hundred days and conducted an experiment on herself.

The lights come up. Applause.

HILDE: So, Ylva, you've been experimenting on yourself! What were you hoping to achieve by memorizing the events of one hundred consecutive days?

YLVA: I thought of how magnificent it would be to be able to remember such a large chunk of a year. It would almost be like an archive. But mostly I wanted to be able to remember more of those magical everyday moments. By memorizing the main events of each day, I hoped they would latch on, like the associations in a memory network. That would be the great bonus.

HILDE: But one hundred days! That's a lot of everyday moments! What exactly did you do to make it work?

YLVA: At first I kept a diary, but that didn't work at all. I know it works for Ida Jackson; maybe she has a very good memory. I don't. I seldom remember what I did last weekend, even if I write it down. So I thought

I would do it like Oddbjørn By: envision a spectacular image for each day. By's method for memorizing a deck of cards in the right order, all fifty-two of them, involves imagining each single card as a distinct image. I only needed thirty-one images, one for each day of the month. It took some time to come up with this, because it had never been done before in memory research. People have memorized decks of cards and the periodic table, but never one hundred days of their own life.

HILDE: Some researchers think that people who are depressed should try to memorize the positive moments of their lives. Since they have trouble remembering these moments, and their memories are often very general, they could use the memory palace technique to remember the good stuff. Is that what you mean?

YLVA: Not quite. For remembering past episodes and keeping them for later, it's all about memorizing events using the method of loci (the memory palace technique), where you place each image along a route. But I didn't use this method because the order for my images is already determined by the dates. So I made up a meaningful image which I attached to each date. I simply manipulated my memories; I was almost creating false memories. The first of the month, as an example, is a lamppost because it resembles the number one and is a distinct object. Then I placed it in my memory even if I hadn't encountered a lamppost on that day in real life.

HILDE: Are there a bunch of people and things hanging out around your lamppost, then? There's a first day in every month, and you memorized one hundred days, so there was more than one first to remember. How did you separate them?

YLVA: I used my real memories, too, and got some help from related events. We both remember the interview with Asbjørn Rachlew in March; it is somehow connected to the rest of the events that month. I imagined an elephant in the conference room, and it now signifies the eleventh of the month. But I also remember what else happened in March and can navigate according to that, even though I have two other elephants to keep track of. Well, I also have a fourth elephant that snuck in. It was the eleventh of May and the experiment was actually over. I was jogging at Ekeberg and became spellbound by the beautiful sunset over Oslo—the same view as in Edvard Munch's painting *The Scream*, actually. I wished I could remember it forever! And then, out of nowhere, there was an elephant in a treetop. I mean—try *not* to think of an elephant! It's impossible!

HILDE: That was perhaps not the beautiful and poetic moment you wanted to remember! So you remember your days by placing a rather vivid image in a memory. Swans, Princess Leia, and tigers, for example. These images don't have an obvious significance, do they?

YLVA: I make a logical association between the date of the month and the image. The fourth is Princess Leia because I like *Star Wars*, and there's a well-known play on words—"*May the fourth* be with you," based on the line "May the force be with you."

HILDE: Doesn't the image mess up the memory?

YLVA: That's the nature of memory reconstruction. There are more layers whenever I look back. One is the memory itself. The other layer is an imaginary world, where the swan or the elephant belongs. Then the memory also has a semantic component, the story of what happened when we interviewed Asbjørn Rachlew and his daughter was sitting with us, for example. Other episodes may also attach themselves to the memory, like running into my friend Gro on the way there. But why are we talking about everything I remember from these hundred days? We were supposed to be talking about forgetting!

HILDE: I'm just thinking that all this memorizing… perhaps it says more about forgetting than remembering.

YLVA: What do you mean?

HILDE: I mean, don't you get a little bit tired of remembering all this stuff? Don't you wish you could forget a lot of it, as you would normally do?

YLVA: Hmm, yes, it's crazy to think, "Oh no! I don't remember the third of March; I have *lost the third of March!*" and feel the same panic as when I put a hand in my pocket and realize my wallet isn't there. I thought that remembering this much meant I was taking control of my memories, but maybe it's the other way around— the memories have taken control of *me*. It becomes more apparent to me how prominent the act of forgetting is in everyday life, especially when you contrast it with this manic struggle to remember it all. Normally, I wouldn't have cared if I didn't remember one day, but suddenly it is so important.

HILDE: Has your conception of time changed, now that you suddenly remember so much?

YLVA: It has given a bit more structure to my life. Since I can remember one hundred days in order, it's like I've captured a sequence of my life, rescued it from transience in a way. As if life's account book has memories on the plus side and things forgotten on the minus side.

HILDE: But it's also similar to when a tree falls in the forest and nobody hears it! If a day has passed and nobody remembers it, it still happened.

YLVA: But still, it can be really frustrating not to remember what happened last fall, for example. I find myself thinking, "When I was on holiday last fall, what did I actually do?" And I draw a blank. It's scary, in a way. Even though my memory is quite normal.

HILDE: Maybe the forgetting is an illusion, because when we talk and get onto a specific subject—running, for instance—you do remember that you ran a half marathon last fall!

YLVA: Yes, that's not easy to forget! As a rule, there are many events we don't forget; they just won't appear at the moment you summon them. Memory processes the event and places it in our life script, and makes it context dependent.

HILDE: After a year, when you haven't rehearsed the hundred days in a while, what will happen?

YLVA: That's the exciting part; I have, after all, "Ebbinghaused," so what will remain after a year? Will I beat the forgetting curve? We will see. But how much of the last hundred days will *you* remember?

HILDE: Hmm, I think I'll remember a fair amount. The most important things, and the ones with the greatest emotional impact, I'll probably remember those best. I don't remember things in order, which you have forced yourself to do. But the important stuff stays. And a lot of what I remember should really be forgotten. Buttering a slice of bread one boring day in February, what's the use of remembering that?

YLVA: I could easily forget the days when I just hung around. Much of what we experience is supposed to get absorbed into the general atmosphere of life. Remembering one hundred days in a row really proves how much we actually forget, despite memory techniques.

HILDE: And if you were to remember every single little detail, it would take a hundred days to remember the hundred days! What's the point of that?

YLVA: I have no need to go back and relive the hundred days exactly as they were. It was really difficult to remember the day I was at home sick, or when I was at home on a Sunday, doing nothing. Or the days we sat at our local café working on this book, like we had done so many times before. According to the laws of memory, those moments are supposed to team up and become a cumulative memory—"writing" or "being sick" or "doing nothing." But I do think that my recollection technique now and then improved things. The seventh of the month was supposed to be the golden day, because my own mild synesthesia makes the number seven sound gold to me. On April seventh it rained, and even though it was a fairly boring and gray day, it rained gold in my mind!

HILDE: As if you were a memory alchemist and transformed the day into gold!

YLVA: It was actually a bit magical. It made my day. But I also appreciate remembering the parts of that memory that didn't include gold. That's what everyday magic really is: knowing that you're alive, letting it rain. But, of course, making up cues is a forced way to access memories. Normally, memories appear via natural association, when we talk about things or hear music.

HILDE: Now the experiment is over. How does it feel?

YLVA: It's wonderful not having to label events all the time. Finally, I can begin to live in the present. Even if it is a bit hard to let go. It is May fifteenth today, and fifteen is always a seahorse. There were a bunch of seahorses hanging by their tails from the beautiful cherry tree outside your door when I arrived.

HILDE: Oh, how symbolic!

YLVA: Yes, but now... it will be such a relief to forget.

The audience applauds. The credits roll. Hilde throws her cue cards over her shoulder and looks knowingly into the camera.

We conducted this little experiment of trying to remember one hundred days to have some fun with forgetting. But for many, forgetting is no fun. People suffer from memory problems in all stages of life and for many different reasons.

One common disease that affects memory is, perhaps surprisingly, depression. Lots of people who suffer from depression worry that their memory is bad. Worrying is a natural part of being depressed; there is so much to worry about when you feel down. You doubt your own abilities. As we know, remembering is characterized by an enormous amount of forgetting and daily mistakes, and this is completely normal. But when you are depressed, you notice only the negative. The glass is seen as half empty, and you believe that you forget differently from happy and optimistic people, who blindly trust what they think of as their own infallible memory. To fill your working memory with worries also limits the space for other things. "I'm worried that I can't remember things" takes the place of "Remember to call Gerda."

Psychology professor Åsa Hammar at the University of Bergen knows very well how depression can tamper with memory. She has tested many depressed individuals and found that they have a normal learning capacity when they're given several attempts to memorize lists of words. But they struggle to remember them after having heard them only once. When the words are repeated, they remember normally. It's as if they're overwhelmed by the first round. Part of the explanation for memory problems associated with depression has to do with attention and working memory, not the actual storing of the memories.

"That's the way it is with most things we try to remember on a daily basis," Hammar says. "Usually, we have only one chance to catch a message." Friends tell us what they did during their vacation just once. We have that one chance to consolidate the information into long-term memory. It isn't strange, then, that individuals with depression feel forgetful, since they need repetition to remember.

"Generally speaking, patients who are depressed also struggle to remember in the aftermath, when they no longer are depressed. They don't remember messages, what to buy in the store, and they miss central elements of conversations. Many may be afraid that they have some brain damage. My studies show, however, that depressed individuals remember as well as others; they just have to allow themselves more time and more attempts to do so."

In collaboration with Yale University, Hammar's research team has discovered another effect of depression on working memory. Individuals who had been depressed were shown several pictures of faces in a row and then got to see one of them again. They were supposed to say where in the row that picture had appeared. They were asked to do this with rows of sad faces and rows of happy faces. This is a relatively simple task, but those with depression struggled disproportionately with the task when the faces were smiling. It was as if they didn't "see" the happy faces. Hammar's explanation is that depressed people have a tendency to be drawn to the negative and almost overlook the positive. Then the complexity of the task was increased, and the subjects had to point out where in the row the face was shown—if the rows were in the reverse order. Their ability to remember the sad faces better made them perform poorly—with both happy and sad faces.

"The bias toward the negative made the sad faces take up more space in working memory, so that they weren't able to implement a reversal of the row," Hammar explains. The effect of bias is clear when people are presented with a difficult task. A particularly interesting finding is that difficulty performing this sort of task can help predict who is at risk of recurring depression. The more difficulty someone has, the greater their risk of recurring depressions. These results suggest that this failure of working memory may be a vulnerability that people with depression unfortunately carry with them, making it harder to keep depressions away.

An incredible number of people, upwards of 12 percent

of the population, suffer from depression, which consequently limits their memory. One percent of the population is affected by epilepsy, making it one of the most common neurological disorders. Henry Molaison had epilepsy, but epilepsy wasn't the main cause of his poor memory; it was surgery that left him with such severe amnesia. Epilepsy itself can cause milder memory problems, though, both in children and adults. Epilepsy is a disturbance in the way the brain functions. Epileptic seizures are caused by uncontrolled electrical activity in the brain; it's like an electrical storm. During a major seizure, a person, who is often unconscious, experiences violent spasms in their arms and legs. The seizure usually doesn't last more than a couple of minutes. But there are other types of seizures, depending on the type of epilepsy. Some have seizures that are so short they are almost undetectable; people just get a distant look in their eyes for a few seconds. These are called absence seizures, because people are "absent" for a few seconds. As mentioned, Henry suffered from these absence seizures as well as major seizures. Even if the seizures last for no more than twenty seconds, they can be enough to disrupt attention and memory formation. Many with absence seizures have attention difficulties extending beyond the seizures, which can make it more difficult for them to learn things at school. Both major seizures and absence seizures are accompanied by epileptic activity in large parts of the brain.

Epilepsy can also stem from a more defined region of the brain. One type of such focal epilepsy is temporal lobe epilepsy, which begins in the temporal lobe. That's where the hippocampus is too. Here it is sometimes a dysfunction, a disruption in normal wiring of neurons, or an injury that causes a focal epileptic seizure. During these seizures, people often feel a sinking feeling in their stomach or a strong feeling of déjà vu, the feeling we all sometimes have that what we're experiencing now has happened already. The only difference is that these feelings of déjà vu are stronger and more frequent in people with epilepsy. Following the déjà vu or the strange gut feeling, the seizure may spread to a wider area of the brain, causing the person to seem "absent" for several minutes, often making smacking noises with their mouth and fidgeting with their hands. Since this type of epilepsy is often caused by damage or dysfunction in the hippocampus, it can also be accompanied by everyday memory problems. People may also forget their seizures, sometimes even the period before and after. Even today, temporal lobe resection is offered as a surgical treatment to some of these people—but only on one side. As long as we have at least one hippocampus left, our memories are safe, relatively speaking.

One person who has tried this surgery is Terese Thue Lund. After dealing with epilepsy for years, without much success with medication, doctors suggested brain surgery. In 2015, surgeons removed about the front inch of her right temporal lobe, including most of her hippocampus, in the hope that it would cure her epileptic seizures.

The jury is still out on the result of the surgery for her epilepsy, but one thing is clear—Terese's memory, which was bad before the surgery, has not been made worse. If anything, it may be better.

If you didn't know, you wouldn't suspect that anything was missing in her brain. We are visiting her at her apartment in Oslo, and she is friendly and hospitable, laughing easily. Her living room is spotless, and on the table are home-baked cupcakes. She tells us she is busy planning her wedding. There is a lot to remember, but she doesn't want it to be stiff and formal.

"I obviously know that I forget things. I have to make a note of all appointments, for example," she begins.

"All of us have to do that."

"Yes, but I have to check my book at least three times a day, and I still don't remember my appointments. When I take the bus, I don't remember what stop I'm getting off at. I never remember people I've met, unless they're wearing unusual clothes or have a strange hair color. I wish people could wear the same clothes all the time!"

"Will you remember us?"

"I could easily walk right by you in the street. I didn't recognize you, Ylva, but I remember now that we had a very good conversation before the surgery. It was in the white building at the hospital...?"

"My office is not in the white building. That's where the occupational therapists and social workers are. Perhaps you talked to a social worker?"

"Oh, I see. Sorry, then it probably wasn't *you* I had that nice conversation with, at least not the conversation that I remember!"

Even though Terese deals with her memory problems with a lot of humor and laughter, it's still difficult for her.

"The worst thing is when people bring up things we've experienced together. Most people know what I'm like, but the fact that I can't remember shared experiences with friends hurts them. My maid of honor remembers all the parties we've been to, in detail. And I remember nothing, although I know we've had fun!"

Terese does not remember her first date with her boyfriend. She doesn't remember if she vacationed with him last year or the year before. She worries a bit about her wedding speech, because she won't be able to charm the audience with stories of the sweet, romantic moments she has shared with him. Her surgery was in December, and over Christmas that year she was convalescent. She remembers that.

"After the surgery, it felt as if the fog lifted a bit. I clearly remember spending Christmas at my in-laws'. I remember that they poked their heads in my room and told me the weather was awful. And I rejoiced! They drove me to the sea, and I stood there and got soaked! I stood there and breathed it all in. With the cold rain on my face I just feel so alive; it's fantastic!"

Terese loves weather, she says, and by that she means bad weather. A wild storm, the sea beating the breakwater in Bodø, the northern city where she's from. She doesn't actually remember what she did during her trip south with her boyfriend, but she remembers in detail how they almost froze to death during a winter camping trip once.

"Perhaps discomfort helps you remember things better? The more uncomfortable you are, the better you remember it, it seems!"

She is laughing. We all find the idea of memory therapy based on being uncomfortable a bit amusing.

Terese's memory is seriously damaged, we know—the three of us sitting around her coffee table, where pictures of the most important things in her life sit under a sheet of glass. There are photos of her with her boyfriend, of northern Norway, and of her dog, as well as comic strips. Undergoing brain surgery where you can potentially lose your memory—what little there is of it—has not been easy for her. She has endured many years of examinations and tests, electrodes both outside and inside her head to measure epileptic seizures, MRIs, memory tests. It wasn't till the surgeon was certain she would not suffer a great further loss of memory that he felt confident enough to cut into her temporal lobe to remove the seahorse and the surrounding brain tissue.

Terese can't study because she forgets everything she reads. She is on a leave of absence from work and has to organize her day in a way that is meaningful to her. She works out, walks the dog, plans her wedding, and meets up with friends. She was diagnosed with epilepsy in 2008. But doctors believe she has had epilepsy since she was a child, when she had nocturnal seizures.

"I'll still have a good life, even though there are many things I can't do. I have a future with a husband and kids before me, and friends and family," Terese says while gently petting her dog, Prudence.

It will be a few years before she knows for sure that her surgery was a success. If she has no more seizures, she can gradually stop taking medication. Epilepsy drugs can also hinder memory as they have a bit of a slowing effect on the brain. Many people have to take several different epilepsy drugs to keep their seizures at bay, while at the same time struggling with side effects. But the alternative—not taking medicine—can also damage memory, especially if the seizures are severe and frequent.

Epilepsy, ADHD, and depression are some of the most common disorders threatening memory from within. But memory can also be damaged from the outside. Head injuries, like the ones you can get in traffic accidents or as a result of sports trauma, are among the most serious threats to the brain, and they mostly affect young people. Falls and accidents can happen to people of all ages, but while older people experience stroke and dementia as natural conse-quences of the aging brain, young people pretty much have only head injuries to worry about, at least in regard to memory loss. Head injuries often lead to memory difficulties. While ADHD affects memory by way of derailing our attention, and temporal lobe epilepsy causes memory loss by damaging the hippocampus, head injuries attack memory from many directions. Attention, working memory, storage, and recall can all be affected, to a greater or lesser degree. On top of all that, many head injury sufferers experience fatigue. They get worn out easily and therefore have diffi-culties staying focused. Even though many head injuries are isolated incidents after which victims improve during the first couple of years, there are also many that cause permanently impaired memory. Head injury is a chronic disorder, even if it was initially caused by an acute incident.

When the memory of a young person is affected, it often happens unexpectedly, because we take memory for granted. Then old age sneaks up on us, and we forget more and more. Throughout our adult life, the cortex shrinks a tiny bit for every year that passes. When we reach old age, it shrinks faster. Brain cavities gradually grow as the brain's white matter disappears. For most people, that's all that happens. We have a difficult time learning new things, and things that are easy to forget, like names, go missing more often than before. But one thing that aging doesn't diminish is the wisdom we have accumulated over a lifetime. Our memories and life experience, even if things gradually take longer to assimilate, become part of one large knowledge bank. Young people may think faster, learn faster, and have a more efficient memory, but old people have an advantage in life experience. Growing old is not about decay, but about change.

With age comes a greater risk for developing brain diseases. The most feared variety of forgetfulness in the present day is Alzheimer's disease. The front pages of major newspapers often report new, small breakthroughs in research. It is one of our era's greatest health challenges, and the hunt for an answer is as complex as the search for a cure for cancer. Julianne Moore won the 2015 Oscar for Actress in a Leading Role for *Still Alice*, in which she portrays a woman afflicted by early-onset Alzheimer's. The desperation the character, a world-renowned linguist, feels when she realizes that she may forget her own children is something very familiar. We could write a whole book on the subject: what happens in the brain, how the disease is experienced by the afflicted person and their family. How nursing homes help by playing music from the patient's youth, awakening something in their reminiscence bump. We cannot do justice to the subject of Alzheimer's disease in a book that's supposed to contain so much else. We have to make do with a couple of paragraphs.

As we live longer than in times past, the task of maintaining the structure and function of our bodies grows bigger. The structure in this case is the brain. Getting wrinkles and liver spots, needing a walker, having a hunched back, losing muscle mass—those are things we can live with. But to lose our memory, and thus lose our grip on existence, is scary. It sneaks up on us. In the beginning, it's difficult to remember names, messages, what we did yesterday. But this is quite similar to what everyone experiences as they age. Memory loss comes with age. It's easy to chuckle when old people start to become a bit forgetful. However, at a certain point it turns into something more serious. As the disease spreads to the whole brain, we need help with everything, and we grow more and more distant. Before it gets that far, though, we've experienced a gradual loss of memory. The hippocampus is affected first, which means that new memories can't be consolidated the way they were before. Alzheimer's patients can tell detailed stories from their own childhood and youth but won't remember that you visited last week. It is a little like the amnesia Henry Molaison had, only not as bad—to begin with. Before we reach the stage of amnesia Henry had, large parts of the brain are affected. In addition to memory problems, Alzheimer's patients struggle with a range of symptoms including language difficulties, emotional disorders, and problems making plans.

No one knows the exact cause of the disease. Some *think* they have found it, while other researchers disagree. So far, the most popular explanation is that waste accumulates around neurons, creating what they call an *amyloid plaque*, and disrupts neurons to the point that they commit suicide (neuronal suicide, that is). This is, of course, not good news for the brain. We all lose a bunch of neurons on a daily basis, but in Alzheimer's patients it happens much faster. Both the ability to consolidate new memories and, gradually, the memories themselves, stored in various places around the cortex, dissolve. Another change in a brain with Alzheimer's is an increase in what is called *tau* protein within neurons, leading to a buildup of damage inside the cells. Researchers disagree on whether it is tau or amyloid plaque that causes the disease. Or perhaps there's an undiscovered culprit? But where do the amyloid plaque and tau come from? Is there anything we can do to prevent them from accumulating in the brain? At the moment, we don't have the answers. However, we do know that the process leading to Alzheimer's starts several decades before the disease is detectable. If we have any chance of halting the disease in the future, we probably have to intervene early, long before we're certain that we're going to get Alzheimer's at all. If we're going to find a treatment that stops Alzheimer's disease before it is too late, it is important to understand exactly how it works. This will require continued, massive effort by thousands of researchers around the world.

What's it like not to remember anything? Do we even know what we're forgetting? That's the definition of amnesia: not remembering, and not knowing what we forget. It's a diagnosis assigned to very few people.

Our friend Henry Molaison from chapter 1 suffered from amnesia, maybe the most severe form. Since he was unable to store memories from moment to moment, everything that happened after his surgery was an isolated moment, trapped in the present. He had his life story, but everything that happened to him after the age of twenty-five (including the two years prior to the surgery that were also gone) didn't remain. He had what is called *anterograde amnesia*. In order to get this form of amnesia, both hippocampi would have to be seriously injured. Damage to other parts of the brain can also result in this type of amnesia if those parts are closely linked to the hippocampus. Stroke, encephalitis, and sometimes even severe heart attacks can all impair both hippocampi.

For some—very few—their whole lives are defined by the inability to remember anything, from the day they were born. They have a rare form of hippocampal dysfunction, something which will affect their development and change the course of their lives. This condition is called *developmental amnesia*, since it begins before a child develops into an adult. The cause isn't always known, but in some cases it is blamed on a difficult birth or respiratory problems in premature babies. The hippocampus is very fragile, and a lack of oxygen can affect it especially severely. This form of amnesia is special because those with it aren't completely unable to form memories, like Henry Molaison—they manage to learn a great deal in school, although they need a lot of help. But what they lack is personal memories. An anonymous English patient whom researchers refer to as Jon is among them. He has an IQ of 114, well above average. He is smart, but he *remembers* nothing! He is married and lives as normal a life as possible. He has probably acquired factual knowledge slowly, through repetition and a meaningful context, but he has no memory of going to school to learn anything. He *knows* that he is married, in the way he knows any other fact, but he has no actual memories of the wedding, first meeting his wife, or their first kiss. He doesn't even know what it's like to recall memories— he's like someone born blind, not knowing what it's like to see.

Most people with any type of amnesia retain memories from earlier in life but struggle to consolidate new experiences. But there's a small subset of people who suffer from another form of severe memory impairment: *retrograde amnesia*. For these people, all earlier memories vanish— they're erased from the hard drive, so to speak. This is one of the greatest mysteries when it comes to forgetting. Since our memories are distributed throughout the brain, how can all of them suddenly disappear? It's hard to imagine that every memory from an entire life could be erased. Nobody has been able to explain how it could be caused by brain damage, either. Sometimes individuals with retrograde amnesia have been discovered far away from home, without any knowledge of who they are. Even in Norway, there have been some famous examples. In December 2013, a man was found lying in a snowbank on the side of a street in Oslo. He didn't remember who he was or where he came from, but he understood several eastern European languages and spoke good English, albeit with an eastern European accent. He was bruised, with cuts all over his body, but the police couldn't figure out what had happened to him. There was likely some criminal context. The man was eventually reunited with his family in the Czech Republic, and the family connection was confirmed by DNA analysis.

For some people, a serious psychological reaction may be the reason for their amnesia. It's as if their entire personality is switched off so they can start over again. The way it often begins is that they will go into a daze and then take off on a trip, apparently without a goal or a reason, typically without identification papers. Some regain their memory through therapy. For others, it's as if the label "I," tied to all personal memories, is erased for good. For some, brain injury (such as from cardiac arrest) may be to blame, but it's still a mystery to memory researchers how memories from an entire life can be gone, as if suddenly deleted. Perhaps the hippocampus is the key here too. The hippocampus ties everything together, all our experiences, the places we've been, and the feeling of connectedness between our memories and our self.

On November 28, 2000, Øyvind Aamot (who now goes by the first name Wind) sent an email from China to his mother. This was the last sign family and friends received that he was alive before he was found three weeks later in a village, with ID papers and plane tickets. He remembered nothing from the previous twenty-seven years, his whole life to that point. He didn't remember who he was, where he came from, or anything he had experienced up to that point. A third of an average life span. Most of us have memories from early childhood, but what is Wind's first memory?

"It's not what people think. Many want to believe that I woke up on a train in China without any memories, because it makes sense based on what they know and understand. But I *didn't wake up* in that way, and I don't think it is that easy to turn what I experienced into a linear story," Wind says.

This is what we know: Wind was twenty-seven years old, working as a freelance journalist, and had developed an interest in anthropology. While on a sailing trip around the world, he told his friends he was going to travel into the Chinese mountains to study equestrian nomads. A month later, no one had heard from him.

He can remember being on a train. He knows that he was found unconscious and sent to the doctor in a car. However, it's difficult to establish when he took the train and when he was at the doctor's. Twice he was found unconscious and helped by villagers in the province of Hunan. It's possible that these episodes have something to do with the matter. A diving accident sustained during his sailing trip could also be the cause, perhaps combined with the fact that he had meningitis as a child. Doctors speculated that some sort of poisoning could have caused the amnesia. Or maybe something completely different had happened. Ten psychology specialists examined Wind afterward without finding the answer.

It took a long time before he became aware of himself again and realized the situation he was in. He just followed others; he was in a passive state. He didn't answer when people asked him questions, he didn't reach out to anyone, he didn't know where he was going. When he saw people standing in line, he joined them. When they reached into their pockets, pulled out something, and handed it over, he did the same, at the same counter. Then he was given food, without having any notion about lineups, stores, or money. Twenty-seven years were completely gone and with them all awareness of how the world works. He had become a man without a memory, a person with retrograde amnesia.

It's a very rare state. There are likely only a few hundred people in the world (we don't know exactly how many there are) with this experience: losing all their memories except for language and motor memories.

Unlike Henry Molaison, people with retrograde amnesia can create new memories. Coincidentally, Henry lost the ability to create new memories at age twenty-seven, while Wind started his life from scratch when he was that age. Twenty-seven-year-old Henry's life extended backward in one direction, while Wind's stretches in the other—combined, the two men have memories of a full life.

"I was helped by so many in the beginning, but I didn't understand the concept of helping. It took a long time before I realized what people were doing for me, and then I was filled with enormous gratitude. I cried long and hard then. In the same way, I didn't know what a friend was, but it was a word I heard all the time, so I noticed it and remembered it. It was always when something good was happening... I started to see everybody that reached out a hand or sent me a friendly glance as a friend," Wind says today.

He has learned to live with his first twenty-seven years being gone. He has reconnected with friends and family. The now more than forty-year-old man has smile lines bearing witness to four decades of laughter, but he has only fifteen years of pleasant memories. But maybe this is an oversimplification.

"You are asking if I miss what I don't remember?" he says and laughs. "How would that be possible? Like you, I have gaps in my memory that I fill in, only my gaps are probably a lot bigger than yours. When someone tells me what I have experienced before, I can visualize it, and it evokes emotions. I call these emotional memories. Immediately following my amnesia, I had no contact with these memories."

Wind has picked up these stories about himself and linked them to the part of his memory which is subconscious, but which reminds him of who he is, of his emotions. When a friend told him about the time he threw a cheese sandwich in his face in elementary school, Wind could picture it, and he could recognize the emotional reaction and the humor between them. That's how he has reconstructed a good deal of his past. It is no longer a large, black abyss of nothingness. He has established continuity with the person he was before he lost his memory—a man with a sparkle in his eye.

How can we determine what's actually true when amnesia devours the memories of the original events? Wind Aamot has filled his past with reconstructions of what he has experienced. They are, in a way, false memories about real events. But it doesn't bother him. He has a feeling of continuity, identity, and truth, even though his life up to the age of twenty-seven is a mixture of reproductions from others and voids in his memory. But everything is linked to the emotional core of Wind's personality.

Which of our memories are true and which are not is something we may never know. It doesn't change who we are. The truth about forgetting is that we are forced to live with it, embrace it, and let it do the job of chiseling out the most important things that will stand out like monuments in our memories, even if that means forgetting all those little things we wish we could remember.

THE SEEDS OF SVALBARD

Or: Traveling into the future

Our revels now are ended. These our actors,
As I foretold you, were all spirits and
Are melted into air, into thin air:
And, like the baseless fabric of this vision,
The cloud-capp'd towers, the gorgeous palaces,
The solemn temples, the great globe itself,
Ye all which it inherit, shall dissolve
And, like this insubstantial pageant faded,
Leave not a rack behind. We are such stuff
As dreams are made on, and our little life
Is rounded with a sleep.

WILLIAM SHAKESPEARE,
The Tempest

IKE PART OF a set from a science fiction film, a building protrudes out of a snowy plateau on Svalbard, an archipelago in the Arctic Ocean. The front of the tall, narrow concrete structure is illuminated by a piece of art that sparkles like snow crystals during the day and northern lights during the night. Otherwise, the structure is completely unremarkable, and most days of the year it stands solitary in this majestic landscape. Through the door there's a corridor that leads to three concrete chambers. Inside these the future of the world's food supply rests in little plastic packets containing various seeds: black, yellow, oblong, round, striped, hairy. Here they wait, side by side.

Svalbard's Global Seed Vault was opened in 2008. Inside, the permafrost keeps the temperature at a steady minus eighteen degrees Celsius all year. The seeds are deposited here from countries all over the world in an international effort, operated by the Norwegian government in cooperation with the Global Crop Diversity Trust. All deposits belong to the countries that made them and can be withdrawn at any time. Hundreds of rice and wheat varieties—each country's agrarian heritage—are kept here. While the seasons turn and winter storms rage, while wars are fought on the other side of the globe and temperatures rise, the seeds lie in the cold, quiet concrete, waiting for the future.

It has been nicknamed the Doomsday Vault. Some have, throughout the process of developing the Seed Vault, envisioned Earth's future, the nuclear wars, climate changes, fields where nothing grows because of drought and new invasions of pests. If the worst happens and the continents look like an uninhabited Mars landscape, the seeds can be withdrawn and give humanity new hope. There has actually already been one withdrawal, after seeds from Syria's national seed bank were destroyed in their civil war. Svalbard's Global Seed Vault is not really meant to be saved for doomsday; it's a running backup for all the depositing nations, for the benefit of Earth. Doomsday is not a single event looming in the distance; it's happening now in the form of natural catastrophes and wars. It's coming gradually, so slowly that we hardly even notice it. Climate change and mass migration are continuously changing our world little by little, every day. The Seed Vault itself has recently been affected by melting permafrost.

But where does the changing future begin? Where do new ideas about the future sprout? In our own seed bank, our memory.

All of us reminisce, more often as we age. Maybe it starts in our twenties, as we read old assignments from elementary school, and ends in a deck chair outside the old folks' home, a photo album in our lap. But reminiscing itself has no evolutionary function. Our flexible, unreliable memories are changeable for one reason: they are supposed to be used; they are not museum objects. Why would nature invest in such an expansive—albeit deceptive—memory if we weren't supposed to use it for something essential? This is where past meets future. One wouldn't be possible without the other. They lie at opposite ends of the dial of our internal time machine. Turn it to the left, and you travel backward in time. Turn it to the right, and you travel forward. Our memories are the prerequisites for mental time travel into the future, for our plans, dreams, and fantasies. It can't be the prospect of an ever-increasing stack of memories to look back on that makes us yearn for eternal life, or at least a very long one. Rather, it must be the thought of always having a future ahead of us. Visions of the future are a natural part of past memories, not only because the past helps us predict the future, but because the process that gives us vivid memories *is the same* as the one that we use to imagine the future.

Counting future thinking among the core functions of memory has not always been the case within memory research. It wasn't something researchers cared much about until the 2000s. One of the pioneers in this regard is Thomas Suddendorf at the University of Queensland in Australia. He speaks to us via the internet from the other side of the globe. Even this is a form of time travel, where eight o'clock in the morning in Oslo meets four o'clock in the afternoon in Australia. In a way, Thomas is speaking to us from the future—just eight hours into the future, but still.

"During all these years, memory researchers have been preoccupied with how people remember correctly, ignoring the important question: Why do we have a memory in the first place?" he says.

In 1994, he and Michael Corballis submitted a paper on the human capacity to imagine the future to a number of psychology journals and were rejected.

"Finally, we were published in a small journal that hardly anyone read," he tells us. The journal has since folded.

"Memory has traditionally concerned what can be measured as correctly remembered. Naturally, future thinking cannot be measured in that way," he suggests as an explanation for the initial lack of interest.

Ten years later, the tables had turned. *Science* magazine named research into mental time travel and episodic future thinking as one of the scientific breakthroughs of 2007. Suddendorf and Corballis's article, "Mental Time Travel and the Evolution of the Human Mind," published in 1997, is one of the most important cornerstones in today's research into episodic future thinking.

"It's a poor sort of memory that only works backwards," says the White Queen in Lewis Carroll's children's classic *Through the Looking-Glass*. It may be that the Queen was quite right: proper memory works both ways.

According to Thomas Suddendorf, the explanation for our fallible memory lies in the evolution of our species. Throughout the history of human evolution, about six million years of it, our environment has changed and forced adaptations in our genetic material. Natural selection does not only favor physical characteristics like opposable thumbs and upright walking, features that aided survival and reproduction among early humans. The human mind was also shaped by evolution. From the perspective of evolutionary psychology, we must always ask what a particular mental function means to survival and reproduction, if it's something universal to humans, not just a local, cultural variation. We can safely say that memory is universal.

"If, for some reason, it was important to humans to preserve an exact copy of the past, then that is what our memory perhaps would give us. But why would we need a replica of the past? It is the future that matters the most. The future holds potential partners and perils. Most of us have a tendency to remember our successes better than our failures. This biases our self-image. What if this is more advantageous when meeting a new potential partner?"

The evolutionary advantage is the strongest argument in favor of a memory system that houses both past and future mental time travel, according to Suddendorf. Or rather: it is actually an argument in favor of memory being a byproduct of the evolution of future imagination. The past is useful only inasmuch as it helps us predict the future. Our malleable, unpredictable, yet vivid memory would not have evolved had it not been for its usefulness in creating vivid, insightful scenes of the future.

"It's the same for all memory functions. Take Pavlov's dogs as an example. They produced stomach acid and drooled when they thought they'd be fed, and when Pavlov repeatedly rang a bell before he gave them food, the dogs began to drool and produce stomach acid at the sound of the bell. This has been used as a prime example of memory, but isn't it more accurate that the dogs drool in anticipation of the food that's coming in the future?"

Pavlov's dogs didn't have to look into their own future, though; they were more like passive recipients of a new connection between sensory impressions in their brain, called *classical conditioning*—a kind of memory without any conscious awareness or will. This type of learning happens in all members of the animal kingdom, from amoebas to humans. But even this primitive form of memory is the product of a need, in all living beings, to be able to predict the future and thereby ensure survival.

"The human way of creating future scenarios and retrieving vivid memories must have had a huge evolutionary advantage. Having an open-ended and flexible memory system allows for an endless set of possible future scenarios that are being constantly evaluated in our minds," says Suddendorf. His book *The Gap: The Science of What Separates Us from Other Animals* takes us on a journey back in time, to when Earth was populated by early hominins: *Australopithecus* species, *Homo erectus*, *Homo neanderthalensis*, and, eventually, *Homo sapiens*. The traces they have left behind actually offer some clues to how their minds must have worked. The emergence of sophisticated stone tools suggests that they were increasingly capable of not only acquiring food, but preparing themselves for the future. "Consider *Homo erectus*, who lived on Earth from around 1.8 million years ago up until as recently as perhaps 27,000 years ago; they developed a kind of handheld axe, useful for cutting meat, among other things. These are elaborate tools, not for throwing away after use. They have carried them along with them—they were armed, so to speak," Suddendorf says.

Homo erectus envisioned a future need for food and protection from predators. Early hominins were scavengers, not hunters, and their need for weapons was for protection—a future-oriented need as much as anything else. Being prepared for perils that may appear at any time in the future is highly advantageous for a creature in the middle of the food chain. As the early humans became increasingly carnivorous, planning for hunting and storage became useful.

An even more convincing argument for a future-oriented *Homo erectus* mind is the discovery of what seem to have been workshops—places where they practiced making axes. They're a powerful clue to understanding ancient future thinking, according to Suddendorf.

"We have found remnants of a number of stone axes in certain locations, as if individuals gathered there to practice making them and teach each other. Consciously honing skills is also something that has made humans very flexible when it comes to preparing for the future. By learning the *art* of making an ax, *Homo erectus* was reassured that he would always have an ax, even if he lost one!"

Homo erectus would be able to face the future equipped for whatever might happen. The ability to imagine future danger and make an ax to defend oneself against it was one of the first steps toward modern humans' spectacular visionary abilities. Telephones and trains, computers and airplanes: none of these would have existed if we hadn't dreamed of them first.

This ability has accompanied humankind across continents, then thousands of years later through a large-scale industrial revolution, all the way to planning an expedition to Mars. Artists, philosophers, and scientists have imagined helicopters (the universal genius Leonardo da Vinci), robots (the author Karel Čapek), spectacular future cities (the film-maker Fritz Lang), and brain scanning for thoughts (the filmmaker Wim Wenders)—the last not unlike what modern fMRI researchers are approaching today. These dreams have, in many cases, predated the technology by hundreds, and sometimes thousands, of years. Even the ancient Egyptians dreamed of traveling to the Moon.

All humans are visionaries, and the basis for their visions lies in their memory. Our memories are the fuel for our imagination. Imagination, in turn, is the energy that brings memories to life. Remembering is actually imagining what happened. Of course, many of the details of a memory are actually stored somewhere. But the moment the memory enters consciousness, in a wave of reexperiencing, it is already reconstructed—the fragments taken from memory are transformed into a coherent experience and story.

Looking at it this way, there's not much difference between reconstructing something that *has* happened and constructing something that has never happened—or hasn't happened *yet*. As with memories, future thoughts aren't built from completely random details. The more we know about the world—the more experiences we've had—the easier it is to picture them as part of the future. Future scenarios are less detailed and less lifelike the further into the future they get. Today's NASA researchers, planning a Mars expedition in the 2030s, can envision more realistic scenarios when they base their projections on pictures they have seen of Mars's surface. Their vision might combine the photos from previous Mars expeditions and perhaps their own experience of climbing a mountain. In the 1700s and 1800s, people imagined Mars in totally different ways. There was no lack of fantasies about who might live on Mars and what the surface of the planet might look like. Its imagined appearance was based on a few sometimes inaccurate images glimpsed through telescopes. What prior experiences caused people to imagine little green men, we will never know.

The best part of this natural time machine is that access isn't restricted to a few lucky people. All of us have it. You may not have noticed it before, but try to think of how much of your day you spend in the future. Are you thinking about what you will have for dinner later today? Are you looking forward to your vacation in two months? You picture it for a few seconds: the plane ride, the warm Jamaican sun bathing your face, the beach and the waves.

There are some situations where it is easier to imagine the future than others—like when you are about to go on a first date. Never are you as engrossed in the future as during the days before the two of you meet. You plan what to wear, where you're going to meet, how you'll greet your date (with a hug or a handshake), what you'll talk about, what you'll do. Internally, you act out the dialogue between the future you and your possible future partner. Sometimes it can seem so real that your feelings are aroused. The synesthetic Solomon Shereshevsky, the man who could not forget, had an almost overpowering imagination. Once, as a schoolboy, he didn't want to leave bed for school, and so he imagined himself going down to have breakfast and leaving for school. The vision was so realistic that he stayed in bed, actually believing he was already at school. For most of us, though, imagining the future is just a natural part of our daily mind wanderings, part of our stream of consciousness, which can just as easily take us back in time as forward. We are all time travelers, all the time.

As far as our brains are concerned, the past and future are almost the same. Only when we consider what the mental time machine actually provides us, in the form of future time travel, can we truly understand the essence of memory, with all its faults and lies. So how do we research the connection between memory and future imaginings? The usual memory tests don't work. Memorizing lists of words can't measure thoughts of the future. For a long time, the future seemed too hard to control, too subjective for research. Envisioning the future was largely the domain of poetry and literature—it has certainly been the cornerstone of the whole science fiction genre. Not until psychology was revolutionized by technology and the invention of fMRI scanning was mental time travel into the future a real interest to researchers. Just as brain imaging made our personal memories "visible," it also suddenly made it possible to see what happens in the brain when we look forward in time.

The opportunity provided by fMRI made Harvard researchers Daniel Schacter and Donna Rose Addis jump on the bandwagon. In 2007, they published an essay in the journal *Nature* titled "Constructive Memory: The Ghosts of Past and Future," which has become an important reference.

Through several experiments, they have pointed to a striking similarity in brain activity when people reminisce and when they imagine the future. Their volunteers are typically presented with a cue word and asked to retrieve a memory, or imagine something that may happen in the future. As they commence their time travels, a set of overlapping brain regions stand out.

Imagine taking part in one of these experiments. You lie down in the scanner, holding in each hand a button to press as responses. A small cage-like device is placed over your head, with a mirror allowing you to look out through the tunnel of the MRI machine to where instructions are presented on a computer screen. While the machine makes all the possible types of clicks and bumps you can imagine (yes, it is a noisy affair), you see cues on the screen, like "beach," triggering a memory or a plan for the summer ahead. For a few minutes, the big, noisy magnetic tunnel is transformed into your own personal time machine, taking you to your summer house where you put your luggage down, kick off your sandals, and open the dusty curtains and the door to the terrace, letting the warm sea breeze embrace you.

After a couple of rounds back and forth in time, the experiment is over, and the researchers can start tracing the time travels in your brain. What are the internal mechanics of this time machine? The hippocampus seems to be involved, but it's not alone. There is also an area in the front and toward the middle of the brain that appears important. An area farther back, also along the midline, looks to be active—it's probably some kind of network hub. Other areas take part too. Time travels make a distinct pattern in the brain, suggesting a network with its own special function. What surprised the researchers perhaps the most was that this network looks suspiciously similar to the so-called default mode network, which is activated when people are asked to try not to think of anything.

We talked about this default mode in the chapter about personal memories—do you remember? Probably not, so we'll do it again. In most fMRI studies of everything from linguistic understanding to working memory, a resting state is used as a control condition—the baseline state of activity we compare the task activity to. That's how researchers are able to show that, compared to not thinking of anything in particular, we largely activate areas in the outer layer of the frontal lobe and the back of the brain when we solve complicated working memory tasks. Relatively speaking, that is. The whole brain is, after all, active at all times. It is the differences between activities that show how we use different areas of the brain more or less.

The default mode, however, isn't just a blank state. What do we normally do when we think of nothing in particular, when we don't focus on a task? We let our thoughts wander. There is a symphony of the past and the future playing in our heads—yes, in yours too—while we are waiting for our next task. We think about what we will do when we are done with the experiment, what we will do later that evening, what we did the previous weekend, and perhaps something fun that happened on the way to the experiment.

There's a lot of evidence that the default, basic state consists of a free flow of memories and future thoughts. According to calculations, people spend more than half their waking time letting their minds wander between memories and future thoughts; what has happened to them and what *might* happen.

"Just think about it," Suddendorf says, "how similar memories and future scenarios feel in our minds. The phenomenology—that is, the quality of the experience—is almost the same."

With a biological footprint in the brain, so to speak, there is no longer any doubt that episodic foresight can be studied scientifically. But the future is studied in other ways as well. The contents of our future time travels can't be measured by fMRI. They're typically measured using questionnaires or interviews, designed to capture the richness of sensory experience as well as its coherence and vividness; it can be as flat as a news story, or as lively as a real experience.

An interesting feature of both memories and episodic future thinking is that their point of view can vary. Are you the "I," or an observer watching yourself from above? Sometimes, you see the events approximately as you did see them, or will see them: through your own eyes. You see the restaurant table sitting between you and your date, your potential future partner on the other side. Other times, you see yourself from the outside: from afar, you see yourself and your date looking at each other across the table—and this feels more distant.

Sometimes, asking people to describe their future thoughts in detail can reveal how they experience imagining it. Some of the story will concern the narrative, how things will unfold. Other details hint at what they experience thinking about it: the emotions, sensory impressions, and personal evaluations. As with memories, future thinking can be either semantic or episodic, factual elements or experiences. To some extent, we can predict what may happen in the future, without picturing it in vivid detail. "Semantic memory is a much older form of memory, in terms of evolution," Suddendorf tells us.

According to him, animals who cache food remember it in a semantic way, not as vivid episodes. It seems impressive when after a long delay, a bird who has hidden away a larva and a nut chooses to ignore the larva and seek out the nut. It may be argued that the bird has an episodic memory of having hidden the larva at that particular place at a specific time and, because of this, now knows that the larva is no longer edible. But according to Suddendorf, it could just as easily be that the bird can tell if the memory has decayed. It has a memory of the larva being in that location, but that memory feels decayed and so is ignored in favor of the memory of the nut, which the bird knows will be edible in spite of the decayed memory.

Even within the span of a human life, the semantic memory appears before the episodic.

"In my research, I have found a convincing overlap between the emergence of episodic memory and the ability to imagine the future in children."

From around age four, kids can relate their past experiences vividly and in detail, as well as plan for the future. They can talk about specific plans and show that they understand that the future may hold different scenarios than the present, including changes in their own needs and states. For example, they may plan on bringing a favorite teddy bear or blanket in case they need comforting.

"The evidence for a unified system where the past and the future go together stems from brain-imaging studies, studies of the similarities in how they are experienced, and from the fact that they develop in parallel in children," says Suddendorf.

An even more convincing argument comes from people with amnesia. It is evident that people with anterograde amnesia (like Henry Molaison, who couldn't store new memories) have only a vague vision of their own future, despite having memories from before the injury. They do have a past, but they can't use it to see the future. It's as if the engine of the time machine is simply not working, even though the fuel of past memories is there. This clearly demonstrates that the future is more than simply learning from past experiences. In 1985 Canadian psychology professor Endel Tulving described an amnesia patient, N.N., who lacked the ability to envision both memories and future thoughts. He asked his patient about the next day:

> E.T.: "Let's try the question again about the future. What will you be doing tomorrow?"
> (There is a 15-second pause.)
> N.N. smiles faintly, then says, "I don't know."

E.T.: "Do you remember the question?"
N.N.: "About what I'll be doing tomorrow?"
E.T.: "Yes. How would you describe your state of mind
when you try to think about it?"
(A 5-second pause.)
N.N.: "Blank, I guess."

When asked to elaborate on the "blankness," N.N. said,
"It's like being in a room with nothing there and having a
guy tell you to go find a chair, and there's nothing there."

There are exceptions, though. Future thoughts don't
always depend on the hippocampus and don't necessarily
use the same machinery as episodic memories. People with
developmental amnesia, those born without the ability to
create episodic memories, can still imagine the future. You
may remember Arne Schrøder Kvalvik, our friend the musician, who could recall hardly any real memories from his
childhood? He still worries about frightening mishaps that
could befall him or his children, so the ability to envision
the future is not missing for him, at least in a semantic way.

Eleanor Maguire and colleagues speculate that this ability may be due to the brain adjusting to a lack of memories
by changing which network contributes to future thinking.
It's similar to the way that children born with damage to
the brain's language center can still learn to talk, because
the brain moves the language center to the healthy half
of the brain. Those who develop amnesia as adults can't
reshape their brain that drastically; it takes a natural malle-
ability that only the developmental potential of childhood

can offer. Those born with a damaged hippocampus can develop new brain networks to take over the vital ability to plan the future. Why, then, can't other parts of the brain take over creating episodic memories in developmental amnesia patients? The answer must be that the hippocampus holds a unique position in binding experiences together in time, at the exact moment they happen. The future has of course not happened yet, and so hasn't been encoded by the hippocampus.

People with depression also have difficulties envisioning the future. For them, the future isn't just gloomy—it's also blurry. Researcher Mark Williams examined a group of individuals with depression in 1996 and found that both their memories and their future thoughts were very fuzzy and general. They did not contain as many specific details as those of happy people. There are consequences to this lack of detail. Seeing the future can present solutions to problems. Imagining a pleasant get-together with friends could mean reaching out to those friends, breaking the isolation that contributes to depression.

Nobody has spent much time researching how depression affects the mental time machine. Only a few studies have been done since Williams's, and they were done in the 2000s. Williams, in turn, has more recently studied depressed patients' future visions of something far worse— suicide. These thoughts are far from fuzzy in people with depression. On the contrary, they are sometimes experienced in a similar way to flashbacks in patients with PTSD. Williams calls them flash-forwards. He and his research team interviewed previously depressed and suicidal individuals about their perceptions of their own deaths. At their most desperate, their suicidal thoughts were very strong and clear. When correlated with a questionnaire that measured the seriousness of suicidal thoughts—in other words, how imminent potential suicide was—there was an obvious connection with the intensity of the suicidal fantasies. The more lucid the fantasy, the more serious the suicide risk. Williams and his research team urge professionals assessing suicide risk in their patients to focus more on the significance of potentially deadly future thoughts. Within clinical psychology and psychiatry, just like in memory research, the phenomenological aspect has been overlooked.

This neglect may be due to the fact that we undervalue the role of fantasy in determining people's actions. Vividly imagining the future: Isn't that a pointless, self-indulgent activity? Isn't knowing what our options are enough? We think of the future in semantic terms too: we consider the likeliness of possibilities to make predictions about our plans for the weekend, what level of education we might get (at least when we have begun studying), how the climate will probably change. Figuring out what the future may hold is of course useful, but do we need to *feel* it? Does immersing ourselves in the future actually have a function, or is it only a side effect of the memory we have? Suddendorf is convinced that visualizing future scenarios does have a point.

"Imagining something before it happens is like a simulation where we can test how the action will affect us and feel different outcomes before we make our choice. For instance, if I want to take a bone from my dog, I can base my actions on past knowledge of how the dog might react. I want to avoid getting bitten, so I picture different scenarios for myself: Should I throw a cat at it to distract it, should I shoot the dog, or should I simply try to calm it before I go for the bone? All these options have different consequences. It would certainly not be very beneficial to kill the poor dog—simply thinking of it is immoral. But in my mental simulations, I can evaluate all of this. The simulation makes all the little details come to life so that they can be scrutinized."

Suddendorf prefers to call future thoughts *episodic foresight*, to use a direct parallel to episodic memory.

So far, we know very little about how important episodic foresight is to shaping our future behavior. We have to remember that there wasn't much interest in future thinking until around 2007. However, a few studies have shown that the episodic system may have a direct influence on problem-solving and creativity. The Harvard professor we mentioned earlier, Daniel Schacter, showed through experiments that when manipulating people into using their episodic memory system more elaborately, they also performed better on a test measuring creativity.

He and his colleagues assigned people to one of two groups and showed all of them a film, which they had to recall later. One group was interviewed thoroughly about the film's details using the interview techniques of modern police investigations—the investigative interview described in the chapter about false memories, the one Asbjørn Rachlew helped introduce to Norway.

The other group received a math assignment simply to pass the time, and to make sure their brains were equally engaged before the creativity task.

Then, during the creativity task, both groups were asked to come up with as many different usages of common objects as they could. (Try it for yourself: How many different things can you do with a pencil?) Those who had gone through the investigative interview were far better at this than the other group. It's as if the interview set their mental time machine into motion, making it useful for creative problem-solving as well.

Another useful aspect of episodic future thinking is that the future can hold rewards for the actions we consider here and now. Imagine that you've just been working out: you're sweaty, you're short of breath, and your legs feel wonderfully heavy as you sink into your couch with a good conscience. Based on previous experiences with working out, you know that this feeling is a treat. By imagining it before you work out again in the future, you get a taste of it, a preview. That motivates you to go and actually work out. We call it *reinforcement*, those good feelings that motivate us to do more of the same. Such future reinforcements help shape our behavior more than we think. In experiments where people are asked to estimate the attractiveness of a larger, distant reinforcement, as compared to a smaller, more imminent reinforcement, they usually favor the more imminent one. But when they are prompted to imagine the distant reinforcement in vivid detail, the difference between the imminent and distant reinforcements in terms of attractiveness here and now is decreased or even eliminated. That actually makes it easier to postpone an immediate need in favor of a later one. Think of this as the cornerstone of human society and civilization. Some argue that these mental previews of future rewards are what made it possible to develop morality. From an evolutionary perspective, human morality originated from the evolution of an ability to postpone selfish immediate needs in favor of the social reward that is the feeling of being part of the group. Experiencing this reward before committing to the socially appropriate behavior makes it more likely that you will actually stick to it. And being part of the group was certainly highly advantageous, both millions of years ago and now.

This supposed advantage for survival and reproduction isn't in itself proof of anything. That's how evolutionary psychology works: it provides us with speculations and assumptions that may bind clues from the past (bones, stones, cave paintings) and the present (how we seem to use our brains) together in a meaningful way. Thomas Suddendorf goes as far as to claim that mental time travel into the future may help explain the most fundamentally human attribute of all: language.

A vision of the future is most useful when you can share it with others; otherwise you'd need to pursue it alone. Our flexible minds require a flexible communication system. These two needs—seeing the future and communicating it—have gone hand in hand during evolution, together pushing the human mind onward. This hasn't just been a dance for two; there's a third partner, our instinctual need to share and bond with each other. While our closest living relatives, the great apes, strengthen their bonds by picking lice from each other's fur, we humans prefer to gossip. We talk with each other, sharing stories and future ideas. Sharing stories is part of what has kept us together in groups, giving us ample opportunities to share our experiences of the past and thoughts about the future. "We don't know for how long, but certainly humans have had a particular drive to tell each other stories. The oldest recorded story that we know is a cave painting in Lascaux in France, dating back seventeen thousand years. It shows a man lying on his back, in front of a bison. The bison's entrails are spilling out, perhaps as a result of the man's spear, or from a hit from a hairy rhino depicted nearby. We don't know exactly how the story goes, but it is obviously a dramatic one, something that was worth telling," Suddendorf says.

Suddendorf believes such stories are crucial to the human capacity to deal with the future. Through stories, we develop our ability to predict the future and come up with plans to face it.

"Through other people's stories, we learn alternative solutions to common problems in life. Most stories, either traditional fairy tales or modern novels, are about people solving problems. They all have a specific moral: if you do what the main character of the story does, the outcome will be like *this*. It increases your repertoire for solving similar problems in your own life in the future. Learning from each other like this is inherent in us. We don't need to invent the wheel over and over again. But most importantly: we exchange visions of the future."

Psychologist and writer Peder Kjøs believes stories are crucial to living our lives and negotiating alternative life paths. If we feel trapped, we can always go to the library and discover between the covers of books other people's thoughts, feelings, and actions, possible parallel worlds to our own. If we don't want to do that, we can turn on the TV, go to the movies, or read news articles. Stories are such important pillars of our society. How come we're drawn to them so strongly?

"We read novels to be able to imagine other ways to live. Each individual's life and fate is an object of intense interest in our culture," he says.

He thinks the story of the individual hero has been reinforced in this way because these days, people are less attached to religion and an all-encompassing relationship with God. What's sacred is an individual's relationship with other individuals, not their relationship with a deity. Stories of personal destinies become guiding stars in our lives. TV series, movies, blogs, newscasts, Facebook updates, and novels, all the stories from around the world and our world history, become a universe of endless possible ways to lead our lives, each something we can choose or choose not to follow. One thing is for sure: if we can't *imagine* something happening, we can't possibly do our part to *make* it happen. Without the impulse to do something new, nothing new will happen.

Suddendorf also maintains that creative writing is made possible by future thinking.

"Imagining fiction is basically the same as imagining oneself in the future. The simulations you perform in your mind can similarly be simulations of different lives. And you can simulate how your story may relate to other people, how relevant and realistic it may be. It is the same brain processes that are at play when you imagine being another person. The memory theater is actually more like a mind theater, where the show can be about yourself in the past or present, or with someone else in the lead role."

In Greek mythology, the goddess of memory, Mnemosyne, was the mother of the nine Muses, inspirational goddesses who ruled over the arts, including several forms of literature. The idea that our ability to imagine—or create—is closely related to memory is not so new after all.

"Memory is the base for everything I write, I think," Linn Ullmann tells us. "Of course, a memory doesn't qualify as art in itself. What sets literature apart from being a simple retelling of a dream you just had is the way you work with memories to transcend the personal moment here and now. I have no obligation to the truth when I write; my only obligation is to fiction itself. That holds true even when I am basing my stories on personal memories."

Her memories may appear unreliable, like memories in real life, without detracting from the emotional strength of her story. Something new emerges: a peephole into other people's lives and fates, feelings and thoughts.

"In my work with memory, I discovered that it borders on imagination," she says.

In her experience, fragments of memories can build something that we would call fiction. When you read a novel, it activates your mental time machine. This time, though, it takes you into the heads of the characters in the book and transports you to the places where they live.

Someone else who believes in the power of storytelling is futurist Anne Lise Kjær, who provides companies, organizations, and nations (she is just back from a consulting trip with the Icelandic government) with tools for building future scenarios that can be used in long-term strategies and marketing work. Her client list includes, among others, brand names like Sony, IKEA, Disney, and several universities.

"My secret is that I am a good storyteller; I bring the future alive in my clients' minds, so that they can see it for themselves," Kjær says.

With an education in design she acquired in Denmark, she is now the CEO of a large, London-based international firm, Kjær Global. When she explains her work as a futurist to us, it is clear that, for the most part, she uses *language* to build the future scenarios. She assembles her stories using the building blocks of stored knowledge and future probabilities. She helps build a *trend atlas*, a map of values and ideas that together may shape the future. This atlas helps shape the semantics of the future. It's a kind of mental Lego. Using these trend analyses, she helps her clients sketch out their own futures. For instance, a typical question to consider for a future vision is whether people are drawn more toward their leisure time than toward earning more money, or how topics of climate change and mindfulness may influence our choices.

"I work with connecting the dots so that the future may appear. But there are many ways the dots can be combined, so we work with parallel visions simultaneously." She tries to see patterns in what is already there.

"Sometimes you need a wild card to make it all fit. These are scenarios that may not be very likely—like green snow! But sometimes they are needed to bring the vision forward," she says. "The future cannot be seen in a crystal ball. I can only provide a road map," she says. Her ability to generate stories about the future has made her a successful entrepreneur. Like Suddendorf, she knows that we shape our own future by recognizing possibilities and following routes that lead to specific outcomes. A futurist is not a fortune-teller; knowledge about the future is mostly a tool her clients can use to influence their own futures themselves.

While Kjær Global is paid to help clients envision a positive future, others imagine a global future that most wouldn't enjoy. There's even a small group of people who deny that these visions could come true.

The Intergovernmental Panel on Climate Change (IPCC), a scientific body guided by the United Nations, works with our globe's future every single day. Katharine Mach of Stanford University is one of the climate researchers responsible for the panel's fifth report, which was released in 2014. The report outlines humankind's past impact on the planet, and the risks we face in the future.

"Picturing ice floating in the water is the easy part. The ice from the melting poles. That is tangible," she tells us.

And that is part of the challenge. Images of polar bears swimming for their lives in the Arctic, far away from us, don't provoke an immediate fear of the climate crisis. We all picture it, the ice floating on the Arctic waves, polar bears swimming desperately as hunting grounds melt away. But that's Svalbard and the Arctic. How will it look for us, where we live?

"The climate panel works with two alternate future scenarios," Chris Field explains. He was the chair of the second working group of the fifth report. As founder of the Carnegie Institution's Department of Global Ecology and director of the Stanford Woods Institute for the Environment, he is without doubt one of the leading experts on world climate change.

"The first scenario is one of ambitious mitigation. The other is a scenario with continued high emissions."

He admits that climate researchers have, for a long time, been reluctant to offer compelling scenarios of how climate change is likely to impact our lives.

"We haven't talked enough about how the future may unfold, and we have not taken into account the psychological aspects of envisioning these future scenarios."

Doomsday visions are the easiest to imagine. However, they are more paralyzing than helpful for most people. If the planet is going under anyway, what's the point of even trying to stop it now?

"It is frustrating to know that we could have achieved the goal of limiting the global rise in temperature if the global community had reacted sooner," he says.

It is impossible not to share his frustration. It feels to us as if he and the UN climate panel carry the future of the entire planet on their shoulders. Isn't it horribly depressing to stare into the abyss climate change represents? Katharine Mach is a young senior researcher, yet she is already all too familiar with it. When she first learned of the gravity of climate change as a much younger scientist, it affected her dramatically.

"It made a strong impression; it was scary, really. It was like going from zero to one hundred very fast; from not knowing to knowing. Now, after so many years of research behind me, it has grown into more of a passion, something that I turn to with the eager curiosity of a scientist."

In Chris Field's and Katharine Mach's vision for the future, there is no room for Hollywood's idea of dystopia. There is no pretentious score, no George Clooney or Tom Cruise in a hero's role.

"The future has three time frames—the immediate future, the near future (the next decade), and the distant future—all three of which depend on what we do here and now. Our job is to draw all these scenarios," Mach says.

We can already see signs of climate change in the form of drought in certain parts of the world and more extreme weather in general. The recent increase in dramatic forest fires, as seen in British Columbia and California, and the overwhelming flooding of Houston, Texas, caused by hurricanes like nothing we've seen before, are only a couple of notable examples. Even in the safe haven of Norway the weather is changing, with less snow in winter and more stormy weather.

For Field, it is important to always keep an eye on the distant future, even though the changes are most visible in the near future. He has a different perspective than what you see in the doomsday visions, though—one that surprises us. Even though the changing climate will bring failing crops, heightened international conflict, and an increasing number of refugees, he insists there are also reasons to be optimistic.

"The ongoing changes in the climate will inspire development. New solutions for green energy will be pushed forward, bringing positive changes, especially for the world's poorest. Climate change may open up possibilities that will make the world a better place, creating stronger and more vibrant communities. This could be our chance to make some substantial changes," Field says.

"Just imagine what the coastal landscape could look like with buildings that have accommodated to rising sea levels," Mach says, explaining that dikes to keep water at bay could also be incorporated into the city's landscape. They could house residences and offices. We imagine a version of the Netherlands, which famously employs dikes to keep it dry, on steroids. Is this the future?

"We have to picture the future within a specific context, to create visions that can be used for people to provoke actual development," she says.

The future isn't just ice drifting with the current until it melts and becomes one with the rising sea, but also cities built from new thinking.

For the IPCC, conveying optimism has been difficult. Their image is one of our shared guilty conscience; they are Hollywood dystopia personified, dressed in nice suits, holding press conferences, and receiving the Peace Prize medal in Oslo City Hall. But they have a commitment to make us aware of all the possibilities, to inspire change. They urge us to take in the future in smaller portions, each possible to swallow, each inspiring new changes as we move along into the near and distant future. For this is the truth about the future: it is not on the other side of an abyss of time, it is right in front of us all the time, like stepping stones in the river (or rather, like floes of ice in the ocean, if that's what it takes to make you picture it). Seeing two weeks into the future may lead to a path that continues on toward the distant horizon. Businesses may find solutions that are beneficial to them here and now; their leaders may picture the economic rewards, their pride at accomplishing something new, the way they can talk about their business at dinner parties to feel good about themselves. They may not picture the fields of golden barley swaying in the wind, an Earth rescued. Money and personal satisfaction are, after all, what drive most people's intentions. Nevertheless, that first small change may lead to the next stepping stone, and then the next. "We are determined to take into account the psychological future; it really concerns us," Field says, and Mach concurs.

Perhaps grandiose Hollywood films featuring "ambitious mitigation" aren't the right direction to follow, but Field and Mach may be on board with the idea of video games that could allow people to take part in more realistic simulations.

For Thomas Suddendorf, climate change and human evolution are intertwined. Because of our ability to make new tools and envision the future, we have outsmarted *Homo erectus*, the Neanderthals, and a whole array of other hominin species. Our highly developed visionary skills have taken us to the top of the food chain. Now we are about to outsmart ourselves with our own civilization-creating, adaptive mind. "It's our ability to envision the future that got us into this mess; let's hope it can deliver us out of it too. To do this, we must envision alternative versions of the future. We can picture the consequences of our actions, so we'd better take responsibility for them."

To make change possible, the future must be transformed into something tangible and meaningful. It must be relevant. Future ways of saving the planet work only insofar as they are related to people's personal goals and preferences, Suddendorf says. For him, personally, the preservation of the rainforest is highly important. He is an eager supporter of work to sustain the gorillas of West Africa and the orangutans of Indonesia. Working in evolutionary psychology, he has spent a lot of time in the presence of these great apes, both in zoos and in the wild. "We all have different motives. For some people, the extinction of species is alarming. We know that using palm oil from plantations in areas that used to be the forest homes of orangutans will mean the end of orangutans as a direct consequence. And that's upsetting. Others, perhaps, couldn't care less about orangutans—some people don't even believe in man-made climate change. But most people can relate to visible pollution, like smog in the air, garbage all around, that sort of thing. When there is filth all around you, it is hard not to take that seriously. So to reach these people, one must appeal to that aspect of environmental problems. For others, economic incentives are more important. Climate change may impact people's economy directly in various ways. We have to build models of the future that mean something to people personally. Our ability to envision the future is, after all, flexible, so it should be possible to mold it into something that fits with people's motivation."

Knowledge is a seed, simultaneously carrying past experience and future potential. The library of Alexandria was once a seed vault, a carrier of a vast amount of wisdom and knowledge, serving the whole Mediterranean civilization until it burned down some two thousand years ago. A new library now resides in Alexandria, like a bridge between ancient culture and present life. The Bibliotheca Alexandrina was designed by the Norwegian architectural firm Snøhetta.

"I'm a production designer for the future. Wherever I go, I see new possibilities, new spaces," Kjetil Trædal Thorsen tells us. He is one of the founding architects of Snøhetta, a now internationally renowned architectural firm.

To be an architect is to envision future cities—or parts of them, especially public buildings, like libraries, theaters, city halls, and opera houses. How they fit into the landscape and the possibilities they reveal will determine the fate of the area surrounding them. Certain solutions for the climate crisis exist within the architecture of the future: we can design buildings that produce more energy than they use, or that are positioned to optimize the sunlight flowing in through windows, or that inspire new ways of thinking about our civilization. Thorsen is involved in making Chris Field's and Katharine Mach's dreams of a positive future a reality.

In addition to the Bibliotheca Alexandrina, Snøhetta is behind the entry pavilion of the 9/11 Memorial & Museum at Ground Zero in New York, where the World Trade Center towers once stood. They designed the new museum in Lascaux, the place in France where some of the world's oldest and most beautiful cave paintings are hidden in a mountain. Spreading out into the Oslo Fjord, the glacier-like marble building that is the home of the Norwegian National Opera & Ballet invites the whole city to its rooftop, like nothing else in the world. The Snøhetta architects work with our shared memory—memories of places—and they do it with their future-oriented design philosophy.

"Of course we as architects don't decide what the future will be. We're simply projecting our wishes for the possible future onto buildings we create," Thorsen says.

With the opera house in Oslo, it all started with values: the Norwegian ethos of equality, open access to nature and the sea, social democracy, closeness with the environment. The architects talked about how buildings can be either authoritarian or inclusive, about high culture versus popular culture—and then, how they could turn something that is traditionally considered an authoritarian, high-culture institution, as represented by grand opera houses around the world, into something that represents the Norwegian spirit of down-to-earthness. When the building was finished, it represented many of these aspects of openness and usefulness to the general public. They thought of a mountain in the middle of the city. Mountains hold a special place in Norwegian people's minds; they are free for all to experience and represent traditional outdoor activities. And so the opera house stands, rising partly out of the sea, with a continuous white surface taking visitors up to the top in a short walk that may feel like climbing a glacier.

"In order to succeed with a project, we have to hit on something that is already a developing tendency in society—our ideas must resonate with something that's already there. Otherwise the ideas will not develop into a real project. Timing is everything, and the jury and the commissioners must believe in it. Actually, there is a series of particular events that must line up for a building like this to finally be realized," Thorsen explains.

At the beginning of this series of remarkable things are stories and ideas—not a lot of drawings. Echoing the futurist Anne Lise Kjær, Thorsen thinks ideas built with language are the departure point for a successful building.

"Architecture is all about telling stories. We are like authors. If we put our designs down onto paper, it closes any further creativity in the other team members. We need to evoke open possibilities, open images, and ideas. That is why we only use words in the beginning of a process. Sometimes, a stray sentence uttered during a brainstorming session triggers an idea that leads to something great."

Collective visions are fostered at Snøhetta. Kjetil Trædal Thorsen never talks about his achievements; he always refers to "our" achievements.

"It is better being part of the Beatles than being Frank Sinatra. We always work together."

The way architects consider the future they're creating in a social context resonates with what we've heard from Thomas Suddendorf, his idea of language and social unity driving the evolution of the human species, giving us the basis for civilization.

Snøhetta now has offices in New York, San Francisco, Innsbruck, Singapore, Stockholm, and Adelaide, employing more than 180 people from thirty different nations. When they won the competition for the opera house, they were located in a shabby street just behind the city's hospital emergency room and had only thirty employees.

"Every assignment is important to us. But the main project, the vision I stick to at all times, is Snøhetta itself, this office, what it will become," Thorsen says.

Not long ago, something happened that moved the architects to tears, something far beyond what anyone could have envisioned for a building they had worked on. It was the dawn of the Arab Spring. The battle against censorship was being fought in the streets of Cairo. With its special status as an international research and cultural center, the Bibliotheca Alexandrina had been exempt from the strict laws of censorship in Egypt. It now became a symbol of the fight for free speech and human rights. As the demonstrations reached a crescendo, the protesters, holding hands, formed a protective circle around the library.

"It's among the greatest things an architect can experience, when people—not the military, but civilians—protect what we have built. We always do our work with humility. We change people's physical environment, so we have to respect humanity. When we build movie theaters in Saudi Arabia, we do it because we feel architecture can be a tool to facilitate something more democratic than what the country stands for today."

Perhaps a building can be so steeped in democratic values that it can change a totalitarian society? The opera house in Oslo has, without doubt, changed people's relationship with an art that was previously associated with the economic and cultural upper classes.

Snøhetta's main office is situated in an old warehouse on the harbor in Oslo. Inside the old walls, the office is a large and open space. From the ceiling, hundreds of small plastic bags filled with water hang in a circle. They catch the light reflected from the fjord and slowly rotate. For a moment, we can almost imagine a tiny seahorse in each bag.

"We made this cheap version of a chandelier for a party once and decided to keep it. Initially, someone had an idea of putting a goldfish in each of them, but that would really just be unethical," the publicist tells us.

She's as relaxed and enthusiastic as everyone else we've met here. We want to stick around; the architects are all open and curious.

Only a few hundred feet down the harbor, the opera house awaits as we stroll along the waterfront. The water is glimmering, almost blinding us with the rays of sun that hold the promise of upcoming summer. Not many months ago, the same water swallowed our diver friends, as they descended fifty feet to perform our memory experiment. Back then the fjord was black with a layer of gray, reflecting the cloudy sky. Rain was pouring over us, and the wind was almost unbearable. Now that June has begun, it all seems so far away: our paper cups of coffee soaked with rain, the divers in their neoprene suits waving optimistically from the water as they prepared for diving while we were left behind on land, shivering in the February rain. How soon experiences that we expect to stay with us fade into vague memories.

The tall windows of the opera house catch the sunlight. Inside, warm oak covers much of the walls, along with colorfully lit origami-like panels made by the Icelandic artist Olafur Eliasson, the same artist who made the rainbow tunnel on top of AROS in Denmark, in the city where Dorthe Berntsen does her research on autobiographical memory.

The opera house is crawling with tourists and regular Oslo citizens, people wearing jeans or saris, kids listening to music, a father and his daughter taking pictures with a selfie stick on the rooftop, children running and laughing with joy. The fjord embraces it all with its glittering light.

YLVA: So what have you learned about memory, Hilde? I mean, while we've been writing this book.

HILDE: A lot of things have surprised me. I've realized that memory has little to do with identity. Personality tests don't measure how much we remember. At the same time, I feel trapped by my memories. They don't become weaker as I get older. It's more like I'm looking at my memories through a prism. From certain angles, I see different colors, but the original colors never disappear. Things have greater depth now than they did in the past.

YLVA: When you talk about it like that, how can memories *not* have anything to do with who we are?

HILDE: No, that's true. I am my memories too. Then I think of the unimaginable number of moments my life contains. So many unique things that have only happened to me, exactly at this point in history. I can't relive any of my life's moments, and only I have experienced them. It's like I'm carrying a whole galaxy of memories inside me!

YLVA: Yes! There are trillions of galaxies out there in space, and it's the same with our memories. There are an unfathomable number of moments. That's why I wanted to study psychology, to learn more about how our inner universes work.

HILDE: And that's why I wanted to be an author. I believe we're all pretending to be normal, orderly, and rational, when in reality we're guided by thoughts, dreams, and wishes we don't even know we have. I've thought a lot about how memory is the core of how we tell stories; we get our storytelling techniques from memory itself. When I write, I have two modes: either "it used to be a certain way" or "then something new happened." In Hollywood films, they always present an initial situation of normalcy, which gives us a feeling of familiarity—which comes from what we in the book refer to as cumulative memory—before something unbelievable happens, usually after twenty minutes, nowadays even faster. This corresponds to those unique, pivotal moments that stand out in our memory. And then there's foreshadowing and flashbacks; that's what we do all the time in life as well. We travel into the future and into the past. And then there's the cliffhanger, that exciting moment in a story when someone's life is in danger and we won't find out what happens until much later. This is life. We seldom figure out the answer to a riddle until some time has passed. We all live with cliffhangers all the time.

YLVA: Yes, the life story, the existential part of memory. The whole time, we live in the story about ourselves and the world, and *that's* what memory is all about. In fiction, there are metaphors and signs, and we look for those in real life too. We're looking for coherence. After an event has passed, we may hang on to an image that's symbolic for the situation or story.

HILDE: But what have *you* learned? Probably nothing, since you're the memory expert?

YLVA: When I did my hundred days experiment, I finally realized how it feels to be one of my own patients. I had to fight to remember each of those hundred days, which of course is ridiculous compared to the kinds of memory problems my patients have. But the struggle to keep on top of it all, to keep track of all that one has to remember, is quite similar, I imagine.

HILDE: I would happily forget some of the negative experiences in life. I wouldn't mind if they could disappear forever. It's a relief to get rid of certain things, to just let go. Forgetfulness is underrated.

YLVA: Sad memories can be pearls on the necklace too. Things don't get better just because we forget the bad things. But I want to stand up for everyday forgetting. It's nice to just live life and give up trying to remember everything the whole time. I will remember this particular day. I'm not sure I will remember in a week exactly what date it was, but that doesn't matter. People seem to be obsessed with memory, thinking remembering better will make them smarter. I understand the obsession. I myself am passionately devoted to thinking about memory and using it the best way. But there is another side to it too. Worrying about our memory is a symptom of society's ridiculous standards of perfection. Not only are we supposed to look good, we're supposed to think perfectly too. It's okay not to remember everything. Memory cannot be perfect.

HILDE: Imagine if we lived forever. Nothing would be important any longer. No moment would be considered remarkable or unique.

YLVA: True. Then it would be even more obvious how important the future is—the whole future would lie in front of us, eternally. But these unique moments we have talked so enthusiastically about, I don't know if I agree anymore that that's where the magic of life can be found. I don't remember my son's first steps, and when I think about it, that makes me sad. Shouldn't I remember his first steps? At the same time, one of my best memories of him, when he was little, is actually a cumulative memory. It is the memory of all the times I lay beside him in bed and sniffed his hair and snuggled his small body. Why is that memory worth less than the moment when he started to walk?

HILDE: Isn't that memory just as nice?

YLVA: Regardless, it is reconstructed. That one moment, a baby's first steps, is as much a reconstruction as is the sum of many moments. We yearn so much to preserve individual moments: something beautiful and precious that will last forever. That doesn't happen. Maybe that's why you write?

HILDE: The fantastic part about writing is being able to preserve precious moments.

YLVA: Or not-so-precious moments?

HILDE: Yes, those too. When I write, memories become hyperrealistic. They become so vivid that I feel like I can touch them. I believe memories are artifacts, meant to be used—like when I retrieve a happy moment and let it linger in consciousness.

YLVA: There is something I've been thinking about a lot lately: mindfulness has given future thinking a bad name. It's like mind wandering has become something bad, something we must take control of! But wandering freely with our thoughts is very natural. We need time to ponder the past, and we need time to look forward, without forcing it. The natural state of the resting mind is wandering back and forth in time.

WHILE WE CONVERSE, a yacht approaches the opera house. It seems to have come out of nowhere, because all of a sudden it's just there. It's filled to the brim with people dressed to party. They have apparently rented an evening on the fjord with drinks and dancing. The speakers are thumping with a song by the Swedish rapper Timbuktu; the artist is spitting the words out in contempt, describing a world where no one wants to plant any seeds or think of anyone but themselves. The boat makes a turn right in front of the opera house. It's almost as if it's doing it just for us. It's easy to visualize the yacht filled with everyone we have talked with for this book: Edvard Moser and Terje Lømo; Eleanor Maguire, busily planning a year of travel; the taxi drivers of London; epilepsy patient Terese Thue Lund and her little dog, sitting so obediently by her feet; Utøya survivor Adrian Pracon, dreaming of a future as a terror researcher; trauma researcher Ines Blix; psychologist Peder Kjøs; writer Linn Ullmann, preparing her next novel; climate researcher Chris Field, currently trying to prevent the temperature from rising two degrees. We can see our sister

Tonje, who has since quit skydiving; and Wind Aamot, who moved to Austria; Henry Molaison; opera singer Johannes Weisser, rehearsing *Onegin*; and criminal investigator Asbjørn Rachlew and his daughter by the gunwale. All of these people, and many others, who have helped us better understand memory and shared their research, their opinions, and their life stories.

The music keeps pumping its depressing message out into the summer air. Maybe we'd like a different song playing as the credits roll for this book, not Timbuktu's rap hit. And perhaps our memory deceives us. Perhaps it was rather the quiet Beatles song "In My Life" we could hear across the Oslo Fjord that June evening, four British musicians singing about all the beautiful moments that can fill the memory, and of all the people you love and lose during a lifetime.

The boat turns away from us, out into the fjord, toward the sunset. You, the reader, can decide on your own credit-roll music, choose what sounds best. What reminds you of the sea? What music moves you? Retrieve that music from your memory!

The music fades, and the yacht disappears over the hazy summer evening's horizon. All those who have been trapped within the pages of this book: we're letting them go now. They're sailing out into the world and into the future. Everything they know, all the memories and experiences they have, they'll use to shape the world and make it better, different, new.

And now it's your turn.

NOTES

Quotations not cited here come from interviews that the authors personally conducted between November 2015 and May 2016.

Chapter 1

Page 2. The discovery of the hippocampus: Shyamal C. Bir, Sudheer Ambekar, Sunil Kukreja, and Anil Nanda, "Julius Caesar Arantius (Giulio Cesare Aranzi, 1530–1589) and the Hippocampus of the Human Brain: History behind the Discovery," *Journal of Neurosurgery* 122, no. 4 (2015): 971–75, doi: 10.3171/2014.11. JNS132402.

Page 4. Henry Molaison's surgery: William Beecher Scoville and Brenda Milner, "Loss of Recent Memory after Bilateral Hippocampal Lesions," *Journal of Neurology, Neurosurgery, and Psychiatry* 20, no. 1 (1957): 11–21.

Page 6. Scoville's confession: Ibid.

Page 7 and on. Henry's life and the scientific discoveries: Suzanne Corkin, *Permanent Present Tense: The Man with No Memory, and What He Taught the World* (London: Allen Lane/Penguin Books, 2013).

Page 7. Henry's ability to perceive time: Ibid.

Page 8. The maze experiment with Henry: Ibid.

Page 9. "I thought this would be difficult": quote from Henry Molaison reported by Brenda Milner on several occasions, including in "Memory as a Life's Work: An Interview with Brenda Milner," interview by Maria Schamis Turner, The Dana Foundation, March 18, 2010, http://www.dana.org/News/Details.aspx?id=43060.

Page 11. "It's a funny thing... I'm living and you're learning": quote from Henry Molaison in Corkin, *Permanent Present Tense*, 113.

Page 12 and on. Solomon Shereshevsky: Alexander R. Luria, *The Mind of a Mnemonist: A Little Book about a Vast Memory*, trans. Lynn Solotaroff (Cambridge: Harvard University Press, 1968).

Page 15. "We believe that the enormous attention": Jacobo Annese, "Welcome to Project H.M.," Brain Observatory, accessed May 20, 2014, http://brainandsociety.org. For further information on Project H.M. and H.M.'s Brain Web Atlas, see https://www.thebrainobservatory.org/project-hm/.

Page 16. Maguire's research has allowed her to "see" memories (Maguire is a coauthor): Martin J. Chadwick et al., "Decoding Individual Episodic Memory Traces in the Human Hippocampus," *Current Biology* 20, no. 6 (2010): 544–47, doi: 10.1016/j.cub.2010.01.053.

Page 18. The infamous "memory wars," review and further research: Lawrence Patihis et al., "Are the 'Memory Wars' Over? A Scientist-Practitioner Gap in Beliefs about Repressed Memory," *Psychological Science* 25, no. 2 (2014): 519–30. doi: 10.1177/0956797613510718.

Page 18. Some argue that the hippocampus has only temporary hold of our memories: Larry R. Squire, "Memory Systems of the Brain: A Brief History and Current Perspective," *Neurobiology*

of Learning and Memory 82, no. 3 (2004): 171–77, doi: 10.1016/j.
nlm.2004.06.005.

Page 19. Others believe that the hippocampus is active every time
we remember: Morris Moscovitch et al., "Functional Neuroanatomy of Remote Episodic, Semantic and Spatial Memory:
A Unified Account Based on Multiple Trace Theory," *Journal of
Anatomy* 207, no. 1 (2005): 35–66, doi: 10.1111/j.1469-7580.
2005.00421.x.

Page 19. "What memory goes with": William James, *The Principles
of Psychology* (New York: Henry Holt and Company, 1890), 651,
https://archive.org/details/theprinciplesofp01jameuoft.

Page 20. Discovery of how neurons are connected (quite a while
before the polar expeditions): Fridtjof Nansen, *The Structure and
Combination of the Histological Elements of the Central Nervous
System* (Bergen: J. Grieg, 1887).

Chapter 2

Page 22. The original diving experiment: Duncan R. Godden and
Alan D. Baddeley, "Context-Dependent Memory in Two Natural
Environments: On Land and Underwater," *British Journal of Psychology* 66, no. 3 (1975): 325–331, doi: 10.1111/j.2044-8295.1975.
tb01468.x.

Page 25. Only thirty-six of one hundred psychology experiments
were re-created successfully: Open Science Collaboration, "Esti-
mating the Reproducibility of Psychological Science," *Science*
349, no. 6251 (2015): aac4716, doi: 10.1126/science.aac4716.

Page 27. Memory as magic in the 1500s and 1600s: Frances A. Yates,
The Art of Memory (Harmondsworth: Peregrine Books, 1969).

Page 31. Tim Bliss and Terje Lømo's first descriptions of memory
trace in the brain: T.V.P. Bliss and T. Lømo, "Long-Lasting
Potentiation of Synaptic Transmission in the Dentate Area of
the Anaesthetized Rabbit following Stimulation of the Perforant

Path," *Journal of Physiology* 232, no. 2 (1973): 331–56, doi: 10.1113/jphysiol.1973.sp010273.

Page 32. O'Keefe's discovery of place cells in the hippocampus: J. O'Keefe and J. Dostrovsky, "The Hippocampus as a Spatial Map: Preliminary Evidence from Unit Activity in the Freely-Moving Rat," *Brain Research* 34, no. 1 (1971): 171–75, doi: 10.1016/0006-8993(71)90358-1.

Page 33. The discovery of grid cells in the entorhinal cortex, right outside the hippocampus, is described here, among other places (the Mosers are coauthors): Torkel Hafting et al., "Microstructure of a Spatial Map in the Entorhinal Cortex," *Nature* 436, no. 7052 (2005): 801–6, doi: 10.1038/nature03721.

Page 35. California researchers found memories connected in memory networks in the hippocampi of mice: Denise J. Cai et al., "A Shared Neural Ensemble Links Distinct Contextual Memories Encoded Close in Time," *Nature* 534, no. 7605 (2016): 115–18, doi: 10.1038/nature17955.

Page 37. Eleanor Maguire's "mind-reading machine": Chadwick et al., "Decoding Individual Episodic Memory Traces."

Page 38. Memories are not static: Heidi M. Bonnici, Martin J. Chadwick, and Eleanor A. Maguire, "Representations of Recent and Remote Autobiographical Memories in Hippocampal Subfields," *Hippocampus* 23, no. 10 (2013): 849–54, doi: 10.1002/hipo.22155.

Page 42. The story about the elephants Shirley and Jenny who found each other after more than twenty years has been widely reported in the media, by, among others, Sophie Jane Evans, "Elephants REALLY Never Forget," *Mail Online,* March 12, 2014, http://www.dailymail.co.uk/news/article-2579045/Elephants-REALLY-never-forget-How-freed-circus-animals-Shirley-Jenny-locked-trunks-hugged-played-met-time-20-years.html.

Page 45. Slime mold that remembers: Tetsu Saigusa et al., "Amoebae Anticipate Periodic Events," *Physical Review Letters* 100, no. 1 (2008): 018101, doi: 10.1103/PhysRevLett.100.018101.

Page 45. Slime mold solves U-shaped trap problem: Chris R. Reid et al., "Slime Mold Uses an Externalized Spatial 'Memory' to Navigate in Complex Environments," *Proceedings of the National Academy of Sciences of the USA* 109, no. 43 (2012): 17490–94, doi: 10.1073/pnas.1215037109.

Page 47. Henry Molaison had only semantic memories of his own past: Sarah Steinvorth, Brian Levine, and Suzanne Corkin, "Medial Temporal Lobe Structures Are Needed to Re-Experience Remote Autobiographical Memories: Evidence from H.M. and W.R.," *Neuropsychologia* 43, no. 4 (2005): 479–96, doi: 10.1016/j.neuropsychologia.2005.01.001.

Page 50. Medical students remember well outside learning environment: Andrew P. Coveney et al., "Context Dependent Memory in Two Learning Environments: The Tutorial Room and the Operating Theatre," *BMC Medical Education*, 13 (2013): 118, doi: 10.1186/1472-6920-13-118.

Chapter 3

Page 54. Marcel Proust, *Remembrance of Things Past* [later translated as *In Search of Lost Time*] vol. 1, *Swann's Way*, trans. C.K. Scott Moncrieff (New York: Henry Holt and Company, 1922), epub edition at http://www.gutenberg.org/ebooks/7178.

Page 56. *After Life* (original title *Wandafuru raifu*), directed by Hirokazu Kore-eda (Japan, 1998).

Page 57 and on. Dorthe Berntsen sums up the research on personal memories in her book *Erindring* [Recollection] (Aarhus, Denmark: Aarhus Universitetsforlag, 2014). In English, see Dorthe Berntsen and David C. Rubin, eds., *Understanding Autobiographical Memory: Theories and Approaches* (Cambridge: Cambridge University Press, 2012).

Page 57. The reminiscence bump: Ibid.

Page 60. Karl Ove Knausgård, *My Struggle*, 6 vols. (5 vols. published in English to date), trans. Don Bartlett, various publishers.

Originally published as *Min kamp* (Oslo: Forlaget Oktober, 2009–11).

Pages 62–63. Aldrin recalls the Moon landing: Buzz Aldrin and Ken Abraham, *Magnificent Desolation: The Long Journey Home from the Moon* (New York: Harmony Books, 2009), 33–38 passim.

Page 64. How to capture people's spontaneous memories in daily life: Anne S. Rasmussen, Kim B. Johannessen, and Dorthe Berntsen, "Ways of Sampling Voluntary and Involuntary Autobiographical Memories in Daily Life," *Consciousness and Cognition* 30 (2014): 156–68, doi: 10.1016/j.concog.2014.09.008.

Page 65. Murakami describes how music awakens memories: Haruki Murakami, *Norwegian Wood*, trans. Jay Rubin (New York: Vintage Books, 2000), 3.

Page 68. "I wanted to see what would happen": Linn Ullmann, *Unquiet* (New York: W.W. Norton, forthcoming).

Page 69. "*To remember* is to look around": Ullmann, *Unquiet*.

Pages 71–72. Two brothers, each with their perspective of memory: William James, *Principles of Psychology*. And Henry James, *A Small Boy and Others* (1913, Project Gutenberg, EBook #26115, 2008), 2, http://www.gutenberg.org/files/26115/26115-h/26115-h.htm.

Page 72. The hippocampus assembles memory elements like a director of a play: Moscovitch et al., "Functional Neuroanatomy."

Page 74. An overview of fMRI studies of personal memories in the brain: Philippe Fossati, "Imaging Autobiographical Memory," *Dialogues in Clinical Neuroscience* 15, no. 4 (2013): 487–90.

Page 75. "Default mode" in the brain: Randy L. Buckner and Daniel C. Carroll, "Self-Projection and the Brain," *Trends in Cognitive Sciences* 11, no. 2 (2007): 49–57, doi: 10.1016/j.tics.2006.11.004.

Page 75. Pitfalls in connection with fMRI studies, elegantly served up with salmon for dinner: Craig M. Bennett et al., "Neural

Correlates of Interspecies Perspective Taking in the Post-Mortem Atlantic Salmon: An Argument for Multiple Comparisons Correction" (poster presented at the 15th annual Organization for Human Brain Mapping conference, San Francisco, CA, June 2009), http://prefrontal.org/files/posters/Bennett-Salmon-2009.pdf.

The study is summarized in an excellent *Scientific American* blog post, "Ig Nobel Prize in Neuroscience: The Dead Salmon Study," September 25, 2012, http://blogs.scientificamerican.com/scicurious-brain/ignobel-prize-in-neuroscience-the-dead-salmon-study/.

Page 77. Individuals suffering from depression have less distinct personal memories: J. Mark G. Williams et al., "The Specificity of Autobiographical Memory and Imageability of the Future," *Memory and Cognition* 24, no. 1 (1996): 116–25, doi: 10.3758/BF03197278.

Page 77. Turning on positive memories to "treat" depression in mice (Tonegawa is a coauthor): Steve Ramirez et al., "Activating Positive Memory Engrams Suppresses Depression-Like Behaviour," *Nature* 522, no. 7556 (2015): 335–39, doi: 10.1038/nature14514.

Page 78. Emotional images for use in student assignments/experiments about sad (and in other ways emotional) memories: Elise S. Dan-Glauser and Klaus R. Scherer, "The Geneva Effective Picture Database (GAPED): A New 730-Picture Database Focusing on Valence and Normative Significance," *Behavior Research Methods* 43, no. 2 (2011): 468–77, doi: 10.3758/s13428-011-0064-1. The (depressing) material can be downloaded from http://www.affective-sciences.org/home/research/materials-and-online-research/research-material/.

Page 79. Tulving's explanation of semantic versus episodic memory: Endel Tulving, "Episodic and Semantic Memory," in *Organization of Memory*, eds. Endel Tulving and Wayne Donaldson (New York: Academic Press, 1972), 381–402.

Page 80. Susie McKinnon tells how she discovered that her memory was different: Helen Branswell, "Susie McKinnon Can't Form Memories about Events in Her Life," *Huffington Post*, April 28, 2015, http://www.huffingtonpost.ca/2015/04/28/living-with-sdam-woman-has-no-episodic-memory-can-t-relive-events-of-past_n_7161776.html.

Page 80. Susie's state is described scientifically here (Levine is a coauthor): Daniela J. Palombo et al., "Severely Deficient Autobiographical Memory (SDAM) in Healthy Adults: A New Mnemonic Syndrome," *Neuropsychologia* 72 (2015): 105–18, doi: 10.1016/j.neuropsychologia.2015.04.012.

Page 81. Arne Schrøder Kvalvik, *Min fetter Ola og meg: Livet og døden og alt det i mellom* [Me and my cousin Ola: Life, death and everything in between] (Oslo: Kagge Forlag, 2015).

Page 83. Extremely good autobiographical memory is described here: Aurora K.R. LePort et al., "Behavioral and Neuroanatomical Investigation of Highly Superior Autobiographical Memory (HSAM)," *Neurobiology of Learning and Memory* 98, no. 1 (2012): 78–92, doi: 10.1016/j.nlm.2012.05.002.

Page 88 and on. All quotes from Adrian Pracon are from our interview with him. He has also told his story in his book, *Hjertet mot steinen: En overlevendes beretning far Utøya* [Heart against stone: The story of a survivor from Utøya] (Oslo: Cappelen Damm, 2012).

Page 91. Flashbulb memories: Roger Brown and James Kulik, "Flashbulb Memories," *Cognition* 5, no. 1 (1977): 73–99, doi: 10.1016/0010-0277(77)90018-X.

An update on the inconsistency of flashbulb memories in one of the most thorough studies ever conducted on the subject: William Hirst et al., "A Ten-Year Follow-up of a Study of Memory for the Attack of September 11, 2001: Flashbulb Memories and Memories for Flashbulb Events," *Journal of Experimental*

Psychology: General 144, no. 3 (2015), 604–23, doi: 10.1037/
xge0000055.

Page 93. Trauma memories are like ordinary memories, just on max-
imum volume: David C. Rubin, Dorthe Berntsen, and Malene
Klindt Bohni, "A Memory-Based Model of Posttraumatic Stress
Disorder: Evaluating Basic Assumptions underlying the PTSD
Diagnosis," *Psychological Review* 115, no. 4 (2008): 985–1011, doi:
10.1037/a0013397.

Page 94. Questionnaire given to 207 government employees after
the 2011 bombing: Øivind Solberg, Ines Blix, and Trond Heir,
"The Aftermath of Terrorism: Posttraumatic Stress and Functional Impairment after the 2011 Oslo Bombing," *Frontiers in
Psychology* 6 (2015): article 1156, doi: 10.3389/fpsyg.2015.01156.

Page 96. Blix calls it centering: Ines Blix et al., "Posttraumatic
Growth and Centrality of Event: A Longitudinal Study in the
Aftermath of the 2011 Oslo Bombing," *Psychological Trauma:
Theory, Research, Practice, and Policy* 7, no. 1 (2015): 18–23, doi:
10.1037/tra0000006.

Pages 96–97. People with PTSD have smaller-than-average hippocampi, and twins have similar-sized hippocampi even when one
isn't exposed to trauma: Mark W. Gilbertson et al., "Smaller
Hippocampal Volume Predicts Pathologic Vulnerability to
Psychological Trauma," *Nature Neuroscience* 5, no. 11 (2002):
1242–47, doi: 10.1038/nn958.

Page 98. The possibility of worse general memory in the wake of
PTSD: Claire L. Isaac, Delia Cushway, and Gregory V. Jones, "Is
Posttraumatic Stress Disorder Associated with Specific Defi-
cits in Episodic Memory?" *Clinical Psychology Review* 26, no. 8
(2006): 939–55, doi: 10.1016/j.cpr.2005.12.004.

Page 99. Treatment methods for PTSD: Jonathan I. Bisson et al.,
"Psychological Therapies for Chronic Post-Traumatic Stress
Disorder (PTSD) in Adults," *Cochrane Database of Systematic*

Reviews, no. 12 (2013): article CD003388, doi: 10.1002/14651858. CD003388.pub4.

Page 99. *Tetris* as a "vaccination" against traumatic memories: Emily A. Holmes et al., "Key Steps in Developing a Cognitive Vaccine against Traumatic Flashbacks: Visuospatial *Tetris* versus Verbal Pub Quiz," *PLOS ONE* 5, no. 11 (2010): article e13706. doi: 10.1371/journal.pone.0013706.

Page 100. PTSD symptoms in the survivors from the Utøya attack: Petra Filkuková et al., "The Relationship between Posttraumatic Stress Symptoms and Narrative Structure among Adolescent Terrorist-Attack Survivors," *European Journal of Psychotraumatology* 7, no. 1 (2016): article 29551, doi: 10.3402/ejpt.v7.29551.

Chapter 4

Pages 106–7. The False Memory Archive: Selected Submissions, A.R. Hopwood's False Memory Archive Website, https://www.falsememoryarchive.com.

Page 110. Memory is fallible and (re)constructive: Daniel L. Schacter, "The Seven Sins of Memory: Insights from Psychology and Cognitive Neuroscience," *American Psychologist* 54, no. 3 (1999): 182–203, doi: 10.1037/0003-066X.54.3.182.

Page 110. The witness who saw two Oklahoma bombers is mentioned in this article, among other places: Daniel L. Schacter and Donna Rose Addis, "Constructive Memory: The Ghosts of Past and Future," *Nature* 445, no. 27 (2007), doi: 10.1038/445027a.

Page 111. People with good autobiographical memories remember more incorrect details (Elizabeth Loftus is a coauthor): Lawrence Patihis et al., "False Memories in Highly Superior Autobiographical Memory Individuals," *Proceedings of the National Academy of Sciences of the USA* 110, no. 52 (2013): 20947–52, doi: 10.1073/pnas.1314373110.

Page 112. Solomon Shereshevsky's vivid childhood memories: Luria, *Mind of a Mnemonist.*

Page 113. Implanted memories in mice: Gaetan de Lavilléon et al., "Explicit Memory Creation during Sleep Demonstrates a Causal Role of Place Cells in Navigation," *Nature Neuroscience* 18, no. 4 (2015): 493–95, doi: 10.1038/nn.3970.

Page 114. The less pleasant experiment, using optogenetics: Steve Ramirez et al., "Creating a False Memory in the Hippocampus," *Science* 341, no. 6144 (2013): 387–91, doi: 10.1126/science.1239073.

Page 116. Staged robbery on TV: Robert Buckhout, "Nearly 2,000 Witnesses Can Be Wrong," *Bulletin of the Psychonomic Society* 16, no. 4 (1980): 307–10, doi: 10.3758/BF03329551. The TV broadcast itself aired on December 19, 1974.

Page 117. The famous asparagus and its way into the memory of unsuspecting volunteers (Loftus is a coauthor): Cara Laney et al., "Asparagus, a Love Story: Healthier Eating Could Be Just a False Memory Away," *Experimental Psychology* 55, no. 5 (2008): 291–300, doi: 10.1027/1618-3169.55.5.291.

Page 117. False memories about food poisoning by egg salad led to changes in food habits in connection with eggs (Loftus is a coauthor): Elke Geraerts et al., "Lasting False Beliefs and Their Behavioral Consequences," *Psychological Science* 19, no. 8 (2008): 749–53, doi: 10.1111/j.1467-9280.2008.02151.x.

Page 117. Loftus's classic car crash experiment: Elizabeth F. Loftus and John C. Palmer, "Reconstruction of Automobile Destruction: An Example of the Interaction between Language and Memory," *Journal of Verbal Learning and Verbal Behavior* 13, no. 5 (1974): 585–89, doi: 10.1016/S0022-5371(74)80011-3.

Page 118. Loftus made people believe they had once been lost at a shopping mall: Elizabeth F. Loftus and Jacqueline E. Pickrell,

"The Formation of False Memories," *Psychiatric Annals* 25, no. 12 (1995): 720–25, doi: 10.3928/0048-5713-19951201-07.

Page 119. Edgar Allan Poe's hot-air balloon hoax was published in the *New York Sun* April 13, 1844.

Page 119. Aloft in a photoshopped hot-air balloon (Maryanne Garry is a coauthor): Kimberley A. Wade et al., "A Picture Is Worth a Thousand Lies: Using False Photographs to Create False Childhood Memories," *Psychonomic Bulletin and Review* 9, no. 3 (2002): 597–603, doi: /10.3758/BF03196318.

Page 127. A critical review of the existence of false memories after manipulation: Chris R. Brewin and Bernice Andrews, "Creating Memories for False Autobiographical Events in Childhood: A Systematic Review," *Applied Cognitive Psychology* 31, no. 1 (2017): 2–23, doi: 10.1002/acp.3220.

    See also this blog post for a critical review of the critical review! Henry Otgaar, "Why We Disagree with Brewin and Andrews," *Forensische Psychologie Blog*, June 1, 2016, https://fpblog.nl/2016/06/01/why-brewin-and-andrews-are-just-completely-wrong/.

Page 129. How a careless mistake by the police can create false memories (Loftus is a coauthor): Kevin J. Cochrane et al., "Memory Blindness: Altered Memory Reports Lead to Distortions in Eyewitness Memory," *Memory and Cognition* 44, no. 5 (2016): 717–26, doi: 10.3758/s13421-016-0594-y.

Page 130. The Innocence Project: http://www.innocenceproject.org. Referenced in: Elizabeth F. Loftus, "25 Years of Eyewitness Science... Finally Pays Off," *Perspectives on Psychological Science* 8, no. 5 (2013): 556–57, doi: 10.1177/1745691613500995.

Page 130. Svein Magnussen's important overview of witness psychology, with examples from Norwegian jurisprudence: *Vitnepsykologi: Pålitelighet og troverdighet i dagligliv og rettssal* [The psychology of eyewitness testimony: Reliability and credibility in everyday life and the courtroom] (Oslo: Abstrakt Forlag, 2004).

Page 131. Memories of serious traumas seldom appear out of the blue: Gail S. Goodman et al., "A Prospective Study of Memory for Child Sexual Abuse: New Findings Relevant to the Repressed-Memory Controversy," *Psychological Science* 14, no. 2 (2003): 113–18, doi: 10.1111/1467-9280.01428.

Page 133. Elizabeth Loftus (coauthor) deprives people of sleep and makes them "confess": Steven J. Frenda et al., "Sleep Deprivation and False Confessions," *Proceedings of the National Academy of Sciences of the USA* 113, no. 8 (2016): 2047–50, doi: 10.1073/pnas.1521518113.

Page 133. False memories of having committed serious crimes in youth: Julia Shaw and Stephen Porter, "Constructing Rich False Memories of Committing Crime," *Psychological Science* 26, no. 3 (2015): 291–301, doi: 10.1177/0956797614562862.

Page 140. Gisli Gudjonsson discusses false confessions in *The Psychology of Interrogations and Confessions: A Handbook* (West Sussex, UK: Wiley, 2003).

Page 142. Asbjørn Rachlew's doctorate: "Justisfeil ved politiets etterforskning: Noen eksempler og forskningsbaserte mottiltak" [Failures of justice within police work: Some examples and research-based interventions] (PhD diss., University of Oslo, 2009), http://urn.nb.no/URN:NBN:no-23961.

Page 142. The description of the perpetrator is from Rachlew's doctorate.

Page 145. Jørn Lier Horst, *Ordeal*, trans. Anne Bruce (Dingwall, UK: Sandstone Press, 2016).

Page 150. Elizabeth Loftus has delivered a TED talk: "How Reliable Is Your Memory?" TEDGlobal, June 2013, https://www.ted.com/talks/elizabeth_loftus_the_fiction_of_memory.

Chapter 5

Page 152. The great taxi experiment (London cab drivers have different hippocampi from the rest of us): Eleanor A. Maguire et al., "Navigation-Related Structural Change in the Hippocampi of Taxi Drivers," *Proceedings of the National Academy of Sciences of the USA* 97, no. 8 (2000): 4398–403, doi: 10.1073/pnas.070039597.

Page 155. Maguire's Ig Nobel Prize description: "The 2003 Ig Nobel Prize Winners," Improbable Research (website), https://www.improbable.com/ig/winners/#ig2003.

Page 156. The great taxi experiment, part 2 (training for the Knowledge changes the brain): Katherine Woollett and Eleanor A. Maguire, "Acquiring 'The Knowledge' of London's Layout Drives Structural Brain Changes," *Current Biology* 21, no. 24 (2011): 2109–14, doi: 10.1016/j.cub.2011.11.018.

Page 159. Neurons are being "born" in parts of the brains of both mice and men: Peter S. Eriksson et al., "Neurogenesis in the Adult Human Hippocampus," *Nature Medicine* 4, no. 11 (1998): 1313–17, doi: 10.1038/3305.

Page 159. Newborn neurons in the hippocampus: Leonardo Restivo et al., "Development of Adult-Generated Cell Connectivity with Excitatory and Inhibitory Cell Populations in the Hippocampus," *Journal of Neuroscience* 35, no. 29 (2015): 10600–10612, doi: 10.1523/JNEUROSCI.3238-14.2015.

Page 159. Researchers have measured activity in new neurons while rats navigate a maze: Ibid.

Page 160. Retired taxi drivers get "normal" brains again: Katherine Woollett, Hugo J. Spiers, and Eleanor A. Maguire, "Talent in the Taxi: A Model System for Exploring Expertise," *Philosophical Transactions of the Royal Society B: Biological Sciences* 364, no. 1522 (2009): 1407–16, doi: 10.1098/rstb.2008.0288.

Page 162. The chess experiment was conducted for the first time in the 1940s but was not published in English until later: Adriaan

D. de Groot, *Thought and Choice in Chess* (The Hague: Mouton, 1965).

Page 163. The experiment was then repeated under similar conditions, but with more chess champions: William G. Chase and Herbert A. Simon, "Perception in Chess," *Cognitive Psychology* 4, no. 1 (1973): 55–81, doi: 10.1016/0010-0285(73)90004-2.

Page 171. Oddbjørn By explains how we can learn memory techniques in his book *Memo: The Easiest Way to Improve Your Memory*, trans. Håkon By (Double Bay, Australia: Lunchroom Publishing, 2007).

Page 173. The Globe as one big memory machine: Yates, *Art of Memory*.

Page 175. Giulio Camillo and the idea of the memory theater: Yates, *Art of Memory*.

Page 175. Robert Fludd considered the theater to have magical connections with our memories: Yates, *Art of Memory*.

Page 175. "Give me my Romeo": William Shakespeare, *Romeo and Juliet*, in *The Globe Illustrated Shakespeare*, ed. Howard Staunton (New York: Gramercy Books, 1998), 188.

Page 178. Solomon Shereshevsky gradually turned to the method of loci: Luria, *Mind of a Mnemonist*.

Page 181. Memory training benefits older individuals (Fjell and Walhovd are coauthors): Ann-Marie Glasø de Lange et al., "The Effects of Memory Training on Behavioral and Microstructural Plasticity in Young and Older Adults," *Human Brain Mapping* 38, no. 11 (2017): 5666–80, doi: 10.1002/hbm.23756.

Page 182. Changes in seniors' memory skills are also visible in their brains (Fjell and Walhovd are coauthors): Andreas Engvig et al., "Effects of Memory Training on Cortical Thickness in the Elderly," *NeuroImage* 52, no. 4 (2010): 1667–76, doi: 10.1016/j.neuroimage.2010.05.041.

Chapter 6

Page 187 and on. Ebbinghaus's travels into the kingdom of forgetful-ness: Hermann Ebbinghaus, *Über das Gedächtnis* (Leipzig: Verlag von Duncker und Humblot, 1885). Translated by Henry A. Ruger and Clara E. Bussenius as *Memory: A Contribution to Experimental Psychology* (New York: Teachers College, Columbia University, 1913), https://archive.org/details/memorycontributi00ebbiuoft.

Page 189. "We cannot, of course, directly observe their present existence": Ebbinghaus, trans. Ruger and Bussenius, *Memory*, 1.

Page 191. "If we remembered everything": James, *Principles of Psychology*, 680.

Page 194. Even gorillas can avoid being noticed, and then they are not lasting memories: Daniel J. Simons and Christopher F. Chabris, "Gorillas in Our Midst: Sustained Inattentional Blindness for Dynamic Events," *Perception* 28, no. 9 (1999): 1059–74, doi: 10.1068/p281059. The original film clip with the basketball players and the gorilla can be seen here: http://viscog.beckman.illinois.edu/media/ig.html.

Page 196. Baddeley and Hitch's original model for working memory: Alan D. Baddeley and Graham Hitch, "Working Memory," *Psychology of Learning and Motivation* 8 (1974): 47–89, doi: 10.1016/S0079-7421(08)60452-1.

An update is found here: Alan Baddeley, "Working Memory: Theories, Models, and Controversies," *Annual Review of Psychology* 63 (2012): 1–29, doi: 10.1146/annurev-psych-120710-100422.

Page 199. ADHD and working memory: Michelle A. Pievsky and Robert E. McGrath, "The Neurocognitive Profile of Attention-Deficit/Hyperactivity Disorder: A Review of Meta-Analyses," *Archives of Clinical Neuropsychology*, published ahead of print, July 6, 2017, doi: 10.1093/arclin/acx055.

Page 199. Worrying disrupts working memory performance: Nicholas A. Hubbard et al., "The Enduring Effects of Depressive

Thoughts on Working Memory," *Journal of Affective Disorders* 190 (2016): 208–13. doi: 10.1016/j.jad.2015.06.056.

Page 201. Solomon Shereshevsky's attempts to forget: Luria, *Mind of a Mnemonist*.

Page 203. An overview of theoretical perspectives of childhood amnesia: Heather B. Madsen and Jee H. Kim, "Ontogeny of Memory: An Update on 40 Years of Work on Infantile Amnesia," *Behavioural Brain Research* 298, part A (2016): 4–14, doi: 10.1016/j.bbr.2015.07.030.

Page 205. Bauer's theory of how early childhood memories gradually disappear: Patricia J. Bauer, "A Complementary Processes Account of the Development of Childhood Amnesia and a Personal Past," *Psychological Review* 122, no. 2 (2015): 204–31, doi: 10.1037/a0038939.

Page 207. An undeveloped grid cell system, before a child begins to move around independently, may explain childhood forgetfulness: Arthur M. Glenberg and Justin Hayes, "Contribution of Embodiment to Solving the Riddle of Infantile Amnesia," *Frontiers in Psychology* 7 (2016): article 10, doi: 10.3389/fpsyg. 2016.00010.

Page 207. Perineuronal nets and their role in memory development: Renato Frischknecht and Eckart D. Gundelfinger, "The Brain's Extracellular Matrix and Its Role in Synaptic Plasticity," in *Synaptic Plasticity*, eds. Michael R. Kreutz and Carlo Sala, *Advances in Experimental Medicine and Biology* 970 (2012): 153–71, doi: 10.1007/978-3-7091-0932-8_7.

Researcher Sakida Palida explains that the perineuronal net can be part of the explanation for childhood amnesia: Laura Sanders, "Nets Full of Holes Catch Long-Term Memories," *ScienceNews*, October 20, 2015, https://www.sciencenews.org/article/nets-full-holes-catch-long-term-memories.

Page 217. Åsa Hammar's research on depression and memory: Åsa Hammar and Guro Årdal, "Cognitive Functioning in Major Depression—A Summary," *Frontiers in Human Neuroscience* 3 (2009): article 26, doi: 10.3389/neuro.09.026.2009.

Åsa Hammar and Guro Årdal, "Verbal Memory Functioning in Recurrent Depression during Partial Remission and Remission—Brief Report," *Frontiers in Psychology* 4 (2013): article 652, doi: 10.3389/fpsyg.2013.00652.

Page 218. Åsa Hammar and researchers from Yale University have discovered another effect of depression: Forthcoming publication.

Page 219. Epilepsy is one of the most common neurological diseases: Susanne Fauser and Hayrettin Tumani, "Chapter 15—Epilepsy," in *Cerebrospinal Fluid in Neurologic Disorders*, eds. Florian Deisenhammer, Charlotte E. Teunissen, and Hayrettin Tumani, *Handbook of Clinical Neurology* 146, 3rd series (2017): 259–66, doi: 10.1016/B978-0-12-804279-3.00015-0.

Page 224. Memory problems are common after traumatic brain injury: P. Azouvi et al., "Neuropsychology of Traumatic Brain Injury: An Expert Overview," *Revue Neurologique* 173, no. 7–8 (2017): 461–72, doi: 10.1016/j.neurol.2017.07.006.

Page 225. A general overview of Alzheimer's disease and possible causes: Kaj Blennow, Mony J. de Leon, and Henrik Zetterberg, "Alzheimer's Disease," *The Lancet* 368, no. 9533 (2006): 387–403, doi: 10.1016/S0140-6736(06)69113-7.

Page 225. *Still Alice*, directed by Richard Glatzer and Wash Westmoreland (New York: Killer Films, 2014).

Page 227. A critique of the amyloid hypothesis, the dominating theory of what causes Alzheimer's: Anders M. Fjell and Kristine B. Walhovd, "Neuroimaging Results Impose New Views on Alzheimer's Disease—The Role of Amyloid Revised," *Molecular Neurobiology* 45, no. 1 (2012): 153–72, doi: 10.1007/s12035-011-8228-7.

Page 228. Developmental amnesia—to be born without ability to create memories because of dysfunction in the hippocampus: Faraneh Vargha-Khadem, David G. Gadian, and Mortimer Mishkin, "Dissociations in Cognitive Memory: The Syndrome of Developmental Amnesia," *Philosophical Transactions of the Royal Society B: Biological Sciences* 356, no. 1413 (2001): 1435–40, doi: 10.1098/rstb.2001.0951.

Page 229. Severe memory loss, called retrograde amnesia, is sometimes attributed to psychological mechanisms: Angelica Staniloiu and Hans J. Markowitsch, "Dissociative Amnesia," *The Lancet Psychiatry* 1, no. 3 (2014): 226–41, doi: 10.1016/S2215-0366(14)70279-2.

... but can also be blamed on brain injuries, as in this case with cardiac arrest: Michael D. Kopelman and John Morton, "Amnesia in an Actor: Learning and Re-learning of Play Passages Despite Severe Autobiographical Amnesia," *Cortex* 67 (2015): 1–14, doi: 10.1016/j.cortex.2015.03.001.

Page 231. Wind Aamot has contributed to a documentary about his amnesia, *Jakten på hukommelsen* [The hunt for memory], text and direction Thomas Lien (Oslo: Merkur Filmproduction AS, 2009).

Chapter 7

Page 237. The Global Seed Vault has been affected by global warming: Amy B. Wang, "Don't Panic, Humanity's 'Doomsday' Seed Vault Is Probably Still Safe," *Washington Post*, May 20, 2017, https://www.washingtonpost.com/news/energy-environment/wp/2017/05/20/dont-panic-humanitys-doomsday-seed-vault-is-probably-still-safe/?utm_term=.763d10f71a68.

Page 239. The article that formed the basis for research of future thinking: Thomas Suddendorf and Michael C. Corballis, "Mental Time Travel and the Evolution of the Human Mind," *Genetic, Social, and General Psychology Monographs* 123, no. 2 (1997): 133–67.

Page 239. Future thinking named one of the year's scientific breakthroughs: The [*Science*] News Staff, "The Runners-Up," *Science* 318, no. 5858 (2007): 1844–49, doi: 10.1126/science.318.5858.1844a.

Page 239. "It's a poor sort of memory": Lewis Carroll, *Through the Looking-Glass*, chap. 5 (1871; Project Gutenberg, 1991, updated 2016), http://www.gutenberg.org/files/12/12-h/12-h.htm.

Page 241 and on. Thomas Suddendorf, *The Gap: The Science of What Separates Us from Other Animals* (New York: Basic Books, 2013).

Page 244. Solomon Shereshevsky imagined he was really at school: Luria, *Mind of a Mnemonist.*

Page 245. A short and easily read essay about the relationship between memories and future thinking (previously cited in chapter 4): Schacter and Addis, "Constructive Memory."

Page 248. Half the time our minds are wandering off from the present, say these researchers, although with a rather dubious conclusion on happiness: Matthew A. Killingsworth and Daniel T. Gilbert, "A Wandering Mind Is an Unhappy Mind," *Science* 330, no. 6006 (2010): 932, doi: 10.1126/science.1192439.

Page 250. Children develop episodic memory and future thinking parallel to each other: Thomas Suddendorf and Jonathan Redshaw, "The Development of Mental Scenario Building and Episodic Foresight," *Annals of the New York Academy of Sciences* 1296 (2013): 135–53, doi: 10.1111/nyas.12189.

Pages 250–51. Tulving's amnesia patient who couldn't see the future: Endel Tulving, "Memory and Consciousness," *Canadian Psychology* 26, no. 1 (1985): 1–12, doi: 10.1037/h0080017. The quotes are from p. 4.

Page 251. How people with developmental amnesia can still picture the future: Niamh C. Hurley, Eleanor A. Maguire, and Faraneh Vargha-Khadem, "Patient HC with Developmental Amnesia Can

Construct Future Scenarios," *Neuropsychologia* 49, no. 13 (2011): 3620–28, doi: 10.1016/j.neuropsychologia.2011.09.015.

Page 252. People with depression have difficulties envisioning the future: Williams et al., "The Specificity of Autobiographical Memory."

Page 252. Suicide thoughts as future thoughts (Williams is a coauthor): Emily A. Holmes et al., "Imagery about Suicide in Depression—'Flash-forwards,'" *Journal of Behavior Therapy and Experimental Psychiatry* 38, no. 4 (2007): 423–34, doi: 10.1016/j.jbtep.2007.10.004.

Page 254. Influence of episodic thinking on creativity: Kevin P. Madore, Donna Rose Addis, and Daniel L. Schacter, "Creativity and Memory: Effects of an Episodic-Specificity Induction on Divergent Thinking," *Psychological Science* 26, no. 9 (2015): 1461–68, doi: 10.1177/0956797615591863.

Page 256. Thinking about the future in detail makes it easier to choose delayed rewards: Jan Peters and Christian Büchel, "Episodic Future Thinking Reduces Reward Delay Discounting through an Enhancement of Prefrontal-Mediotemporal Interactions," *Neuron* 66, no. 1 (2010): 138–48, doi: 10.1016/j.neuron.2010.03.026.

Page 261. The UN climate panel's fifth report: Chris Field et al., eds., *Climate Change 2014: Impacts, Adaptation, and Vulnerability. Part A: Global and Sectoral Aspects. Contribution of Working Group II to the Fifth Assessment Report of the Intergovernmental Panel on Climate Change* (Cambridge and New York: Cambridge University Press, 2014).

Page 278. "In My Life," performed by the Beatles, songwriters Lennon–McCartney, track 11 on *Rubber Soul*, LP, Parlophone, 1965.

RECIPE FOR
GOOD MEMORIES

Or: Thanks to all involved

BY WRITING THIS book, we have created new memories together. We have in a way given ourselves a new, unique reminiscence bump. For this, we owe a whole lot of people our thanks, because such memories are not created in a vacuum.

When diving into the memory, it is absolutely an advantage to receive help from good divers. "How many divers do you need?" Caterina Cattaneo said when we asked her if she would help us re-create the famous experiment. Caterina, who is also a brilliant author and our good friend and support, has contributed a lot more than diving, but now it's all about diving. We also want to thank Tine Kinn Kvamme, Rune Paulsen from the Divestore at Gylte, and the ten fantastic men who took part that rainy, raw February day.

The four chess champions Simen Agdestein, Olga Dolzhikova, Aryan Tari, and Jon Ludvig Hammer deserve glory and thanks, even if their efforts probably mean less in their personal memory bank than real chess achievements. For the hot-air balloon experiment, we enlisted the help of our Norwegian editor's wife, Anita Reinton Utgård, who benevolently entered into an agreement of conspiracy and provided childhood photos of our skilled editor Erik Møller Solheim, whom we of course also want to thank! Erik knew what we were after from the beginning and has helped us make the best possible book.

We want to thank Adrian Pracon for our most precious memory from this time. He brought us to Utøya and showed us the places that mean so much to him, both the pleasant places and the places where terror attacked him so directly.

Thank you all who have contributed to this book in interviews! Thanks to each and every one of our interviewees, but perhaps particularly to Linn Ullmann, Peder Kjøs, Terese Thue Lund, and Arne Schrøder Kvalvik, who properly changed our lives just a little bit and made us wiser in our work on this book.

Beyond the pages of the book, we have had a large and varied support group, including among others Simon Grahl; Mia Tuft; colleagues and friends from the neuropsychology environment; Hilde's book club and author friends Eivor Vindenes, Tone Holmen, Hedda Klemetzen, and Vera Micaelsen; and fellow hiker Marit Ausland. The largest support group includes our families: Matt and Niclas, Liv, Heidar, and Eyvor.

Although not by means of hot-air balloon, the book has now traveled all the way over to Vancouver, Canada, with the generous help of Rob Sanders and all the great staff at Greystone Books, including Jennifer Croll and Dawn Loewen, and the translator Marianne Lindvall. Together, we made this book even better.

Finally, we want to thank our sister Tonje, not only for her generous sharing of her near-death experience, but for all the memories we share. Like the time we all got stuck in the swamp near our cabin in Bodø: Ylva got stuck first. Tonje came to rescue her and got stuck too. Then Hilde came to the rescue but wasn't exactly lucky either. There we stood, calling our father, who eventually pulled us out. This goes to show that we don't always learn from the immediate past. The experience, though, we can keep with us for the rest of our lives.